SpringerBriefs in Applied Sciences and Technology

SpringerBriefs present concise summaries of cutting-edge research and practical applications across a wide spectrum of fields. Featuring compact volumes of 50–125 pages, the series covers a range of content from professional to academic.

Typical publications can be:

- A timely report of state-of-the art methods
- An introduction to or a manual for the application of mathematical or computer techniques
- A bridge between new research results, as published in journal articles
- A snapshot of a hot or emerging topic
- An in-depth case study
- A presentation of core concepts that students must understand in order to make independent contributions

SpringerBriefs are characterized by fast, global electronic dissemination, standard publishing contracts, standardized manuscript preparation and formatting guidelines, and expedited production schedules.

On the one hand, **SpringerBriefs in Applied Sciences and Technology** are devoted to the publication of fundamentals and applications within the different classical engineering disciplines as well as in interdisciplinary fields that recently emerged between these areas. On the other hand, as the boundary separating fundamental research and applied technology is more and more dissolving, this series is particularly open to trans-disciplinary topics between fundamental science and engineering.

Indexed by EI-Compendex, SCOPUS and Springerlink.

More information about this series at http://www.springer.com/series/8884

Urban Kuhar · Gregor Kosec · Aleš Švigelj

Observability of Power-Distribution Systems

State-Estimation Techniques and Approaches

Springer

Urban Kuhar
Jožef Stefan International
Postgraduate School
Ljubljana, Slovenia

Gregor Kosec🆔
Jožef Stefan Institute
Ljubljana, Slovenia

Aleš Švigelj🆔
Jožef Stefan Institute
Ljubljana, Slovenia

Jožef Stefan International
Postgraduate School
Ljubljana, Slovenia

ISSN 2191-530X   ISSN 2191-5318   (electronic)
SpringerBriefs in Applied Sciences and Technology
ISBN 978-3-030-39475-2   ISBN 978-3-030-39476-9   (eBook)
https://doi.org/10.1007/978-3-030-39476-9

This Springer imprint is published by the registered company Springer Nature Switzerland AG
The registered company address is: Gewerbestrasse 11, 6330 Cham, Switzerland

To all who make our lives worthwhile.

Preface

Electricity has an enormous influence on our daily lives and the economic development of society. As a result, the efficient transmission and distribution of electricity are fundamental to our health and well-being. Primarily because of the growth in distributed energy resources, the flow of electricity from utility to consumer has become a two-way street. With such a change in our power-distribution systems, we will increasingly rely on the real-time observation and control of the electricity system to ensure its safe and reliable operation in the future.

A distribution system's state estimation represents a crucial part of ensuring observability in smart-grid systems. This book describes the design and implementation of a three-phase state estimation that is suitable for power-distribution networks. It gathers all the relevant state-of-the-art knowledge, adds the missing pieces and brings the subject together in such a way that the reader is given a complete picture of the relevant design and implementation factors as well as the methods to address them. The modeling of all the major power-distribution components to enable a three-phase network model's construction is described; sensitivity analyses showing how uncertain conductor lengths influence the state-estimation results are presented; and methods for the numerical solution of different state estimators and their application are reviewed and evaluated. The target audience is engineers and industrial practitioners, on the one hand, and electrical engineering students, researchers, and academics in the field of power systems and distribution systems, on the other.

Some parts of this book were originally in Urban Kuhar's doctoral dissertation, written during his studies at the Jožef Stefan International Postgraduate School (Ljubljana, Slovenia). The research was supported by the EU's SUNSEED

(Sustainable and robust networking for smart electricity distribution) project under Grant FP7 619437 (the European Community's Seventh Framework Programme), and by the Slovenian Research Agency (research core funding No. P2-0016 and P2-0095).

Ljubljana, Slovenia   Urban Kuhar
January 2020   Gregor Kosec
                                                                             Aleš Švigelj

Contents

Acronyms

AMI	Advanced Measurement Infrastructure
API	Application Programming Interface
AWG	American Wire Gauge
BLUE	Best Linear Unbiased Estimator
CRLB	Cramer–Rao Lower Bound
DER	Distributed Energy Resources
DG	Distributed Generation
DSO	Distribution System Operator
DSSE	Distribution System State Estimation
EV	Electric Vehicles
FDIR	Fault Detection, Isolation, and Service Restoration
GMR	Geometric Mean Radius
GUI	Graphical User Interface
IRLS	Iteratively Reweighted Least Squares
JSI	Jožef Stefan Institute
KKT	Karush–Kuhn–Tucker
LAV	Least Absolute Value
LOMED	Low Median
LS	Least Squares
LSE	Least Squares Estimation
MC	Monte Carlo
MLE	Maximum-Likelihood Estimation
MQTT	Message-Queuing Telemetry Transport
MVUE	Minimum-Variance-Unbiased Estimator
PDF	Probability Density Function
PLC	Power-Line Communications
PMU	Phasor Measurement Unit
SE	State Estimation
SG	Smart Grid
SHGM	Schweppe–Huber Generalized-M

SM	Smart Meter
TSO	Transmission System Operator
WLS	Weighted Least Squares

Symbols

Θ_m^a	Voltage phase angle at node m on phase a
$|V_m^a|$	Voltage magnitude at node m on phase a
E_k^a	Bus voltage phasor
I_{km}^a	Line current phasor
P_{km}^a	Active line power flow
Q_{km}^a	Reactive line power flow
S_{km}^a	Complex line power flow
P_k^a	Active bus power injection
Q_k^a	Reactive bus power injection
R	Resistance
X	Reactance
C	Capacitance
z_{km}^{ij}	Conductor series impedance element between buses k and m
r_{km}^{ij}	Conductor series resistance element between buses k and m
x_{km}^{ij}	Conductor series reactance element between buses k and m
$y_{km,sh}^{ij}$	Conductor shunt admittance element between buses k and m
$g_{km,sh}^{ij}$	Conductor shunt conductance element between buses k and m
$b_{km,sh}^{ij}$	Conductor shunt susceptance element between buses k and m
G_{km}^{ij}	System-admittance conductance element
B_{km}^{ij}	System-admittance susceptance element
t_{km}^{ij}	Connection element between buses k and m, on the side of bus k
J	Objective function of the state estimator
J^*	Objective function of the state estimator for the optimal solution
$\hat{z}_{ij}$	Primitive series impedance element
$\hat{P}_{ij}$	Primitive potential coefficient element
f_i	Measurement i
x_i	State variable i
h_i	Measurement function i

r_i	State-estimation residual
c_i	State-estimation equality constraint
g_i	State-estimation inequality constraint
γ_i	SHGM estimator weight
ρ	SHGM estimator function
ξ_i	Number of nonzero entries in the i-th row of the Jacobian matrix
ϵ_0	Permittivity of free space
ϵ_r	Relative permittivity of the medium
p	Probability density function
GMR_i	Geometric mean radius of conductor i in feet
D_{ij}	Distance between conductors i and j in feet
S_{ii}	Distance from conductor i to its image in the ground
S_{ij}	Distance from conductor i to image in the ground of conductor j
RD_i	Radius of conductor i
d_c	Phase conductor diameter
d_{od}	Nominal diameter over the concentric neutrals of the cable
d_s	Diameter of a concentric neutral strand
r_c	Resistance of the phase conductor
r_s	Resistance of a solid neutral strand
k	Number of concentric neutral strands
T	Thickness of copper tape shield
w_i	Noise contribution to measurement i
m	Number of measurements
n	Number of state variables
o	Number of equality constraints
l	Number of inequality constraints
σ_i	Standard deviation of measurement noise contribution i
Ω_p	Set of phases
Ω_I	Set of active inequality constraints
Λ	Set of buses adjacent to bus k including bus k
$\Theta(\lambda, \mu)$	Dual function of the optimization problem
$\kappa(\mathbf{A})$	Condition number operator
α	Hachtel's scaling factor
$\mathscr{L}$	Lagrangian function
$\mathscr{N}$	Normal probability density function
$\mathbf{f}$	Vector of measurements
$\mathbf{g}$	Function vector
$\mathbf{x}$	Vector of state variables
$\mathbf{h}$	Vector of measurement functions
$\mathbf{z}$	Vector of measurements
$\mathbf{r}$	Vector of residuals
$\mathbf{c}$	Vector of equality constraints
$\mathbf{g}$	Vector of inequality constraints

$\mathbf{a}$	Vector of measurements and model data that are used in the calculation of sensitivities
$\mathbf{I}_{km}^{a,b,c}$	Vector of branch currents
$\mathbf{E}_{k}^{a,b,c}$	Vector of bus voltages
$\mathbf{P}_i$	Vector of active power injection measurements in the system
$\mathbf{Q}_i$	Vector of reactive power injection measurements in the system
$\mathbf{P}_{ij}$	Vector of active power flow measurements in the system
$\mathbf{Q}_{ij}$	Vector of reactive power flow measurements in the system
$|\mathbf{V}|$	Vector of voltage magnitude measurements in the system
$\mathbf{\Theta}$	Vector of voltage phase measurements in the system
$\widehat{\mathbf{x}}$	Estimated vector of state variables
$\mathbf{w}$	Vector of noise contributions to measurements
$\boldsymbol{\lambda}$	Vector of Lagrange multipliers for equality constraints
$\boldsymbol{\mu}$	Vector of Lagrange multipliers for inequality constraints
$\mathbf{x}^*$	Vector of state variables at the optimal solution
$\boldsymbol{\lambda}^*$	Vector of Lagrange multipliers for equality constraints at the optimal solution
$\boldsymbol{\mu}^*$	Vector of Lagrange multipliers for inequality constraints at the optimal solution
$\mathbf{R}$	Measurement covariance matrix
$\mathbf{I}$	Identity matrix
$\mathbf{0}$	Null matrix
$\widehat{\mathbf{z}}_{km}$	Primitive series impedance matrix
$\widehat{\mathbf{P}}_{km}$	Primitive potential coefficient matrix
$\mathbf{C}$	Matrix of first order partial derivatives of equality constraints
$\mathbf{C}_{abc}$	Capacitance matrix
$\mathbf{Z}_{km}$	Series impedance matrix between buses k and m
$\mathbf{B}$	System conductance matrix
$\mathbf{G}$	System susceptance matrix
$\mathbf{Y}_{km,sh}$	Shunt admittance matrix between buses k and m
$\mathbf{Y}_{b,km}$	Branch admittance matrix between buses k and m
$\mathbf{T}_{km}$	Connection matrix between buses k and m on the side of bus k
$\mathbf{T}_{mk}$	Connection matrix between buses m and k on the side of bus m
$\nabla_{\mathbf{x}}$	Gradient with respect to vector $\mathbf{x}$
$\nabla_{\mathbf{xa}}$	Hessian with respect to vectors $\mathbf{x}$ and $\mathbf{a}$ respectively

Chapter 1
Introduction

1.1 The Evolution of Power-Distribution Systems and the Need for the Implementation of a State Estimation

As individuals we depend on electricity to heat, cool, and light our homes, to refrigerate and prepare our food, to pump and purify our water, to process our sewage, and to provide most of our communications and entertainment. As a society, we depend on electricity to light our streets, to control the flow of traffic on the roads, rails and in the air, to operate the myriad physical and information supply chains that create, produce, and distribute goods and services, to maintain public safety, and to ensure national security [1].

While our society is becoming ever-more dependent upon electricity, the electricity system is undergoing a complex transformation. In the past 10 years, electricity production and distribution have faced two major changes. First, production is shifting from conventional energy sources, such as coal and nuclear power, to renewable, low-carbon resources, such as solar and wind. Second, consumption in the low-voltage grid is expected to grow significantly due to the introduction of electrical vehicles. Moreover, these changes are being accelerated by government incentives to install renewables, such as wind and solar power plants. Consequently, a large number of small units (in terms of nominal power) have been installed and connected to the distribution grid. This invalidates the traditional view of power systems as consisting of large-scale generation units connected to the transmission grid that supplies distribution networks, which then unidirectionally supply the end customers. In addition, the future will see investments in the domain of electric mobility. Currently, the world's transportation sector accounts for more than a quarter of all energy consumption and is, almost exclusively, fuelled by oil products.

The most natural point of connection for electric vehicles (EVs) is the distribution system. In addition to EVs, large-scale energy-storage devices are expected to be deployed throughout the network. All these changes have the potential to greatly

© The Author(s), under exclusive license to Springer Nature Switzerland AG 2020
U. Kuhar et al., *Observability of Power-Distribution Systems*,
SpringerBriefs in Applied Sciences and Technology,
https://doi.org/10.1007/978-3-030-39476-9_1

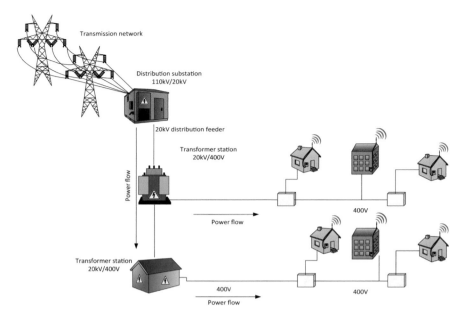

Fig. 1.1 Power-distribution networks in the past

reduce the total carbon footprint. However, they also significantly increase the system's complexity, which means it will have to operate within even stricter reliability margins.

Historically, distribution grids were designed for unidirectional power flow. The assumption was that a part of the distribution network would be connected to the transmission network, and then power flows from that connection point through a medium-voltage feeder into the low-voltage networks and to customers (see Fig. 1.1). Distribution networks were built with a radial topology, with the medium-voltage feeder operating in a disconnected loop, which ensures the quick restoration of power in the case of a single branch outage in that loop. Compared to transmission networks, the cost of a meshed topology would be higher than the costs of an outage or two in the network. A radial topology and the assumption of unidirectional power flow also greatly simplify the design and operation of the network. Loads can be estimated and predicted from urban development plans, and using load-flow calculations, the voltage profile can be determined along the feeder.

In industrialized countries, the typical consumers are expected to become electricity producers as a result of the ongoing widespread deployment of distributed generation and energy-storage elements. Thus, in recent years, a massive increase in distributed generation (DG), such as wind and solar installations, has been witnessed [2]. In most cases, the DG is connected into the distribution network because the nominal powers are relatively small ($\approx 10 - 10^3$ times the power of an average consumer, as opposed to a conventional power plant that is $\approx 10^4 - 10^7$ times the power of an

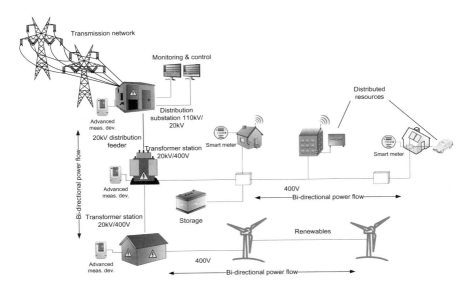

Fig. 1.2 Evolution of power-distribution networks

average customer). Another fundamental change that is underway is the adoption of EVs [3, 4]. Their growth is expected to have a major impact on distribution systems [5], since the power of the inverter and the battery capacity are significant (orders of $\approx 10\,kW$ and ≈ 70 kWh, respectively). Since an EV is essentially a battery that is connected to the distribution grid, it can perform as a load or as a generator. Both DGs and EVs are commonly referred to as distributed energy resources (DERs).

The adoption of DERs is leading to bidirectional power flows in distribution systems, which the systems were simply not designed for. In turn, bidirectional power flows (see Fig. 1.2) can cause under- or over-voltage events, etc. [6, 7]. Since the generation of DERs is irregular and intermittent, it is almost impossible to combine them with conventional electricity generation. As an example, the wind can vary significantly from one time of the day to another, which means that over the course of a few minutes the power output from a wind farm could change dramatically. Solar power can suffer similar problems. For instance, even a small cloud can reduce the output of a photovoltaic solar farm for a few minutes before the output returns to its previous level. This unpredictable generation and the continuously changing behavior of the consumers make power flows uncertain.

The ramp rates of conventional power plants (such as coal and nuclear) are far too inflexible to achieve the rates that are required by renewables [8]. As conventional generators are being shut down, another significant phenomenon is occurring, i.e., the reduction of inertia. In order to ensure a stable supply of electricity, the system must be able to cope with the sudden loss of generation. Battery storage can provide a frequency-control reserve [9–12], but there is still a need for the coordinated control of such assets in the network.

By using only conventional grid-management systems, the capacity of the electrical grid is under question. Reinforcing the grid all the way to the user is an option, albeit an expensive one, especially when other, more convenient and cheaper alternatives are on the horizon. The shift from unidirectional power flow toward a fully bidirectional paradigm can be used as an advantage, allowing the installation of additional DERs within the existing infrastructure.

The described context gave rise to the concept of the smart grid (SG) as a modern electricity network in which advanced, automated functions are implemented, exploiting a suitable communication infrastructure to efficiently perform the management, control, and protection of the network [13]. However, this requires the precise monitoring of the distribution grid that provides reliable and accurate information about its status to enable dynamic grid management in the future [14].

In general, distribution systems still lack observability or monitoring capabilities, despite the fact that they require this because of their considerable diversity and variability, as well as vulnerability to disturbances. To effectively control and manage the assets in a network, an accurate picture of the current operating conditions must be produced. Information that is extracted from the measurement system, which comprises measurement devices deployed across the network, as well as all the processing and collection algorithms and software. These devices are usually referred to as phasor measurement units (PMU) and are normally deployed over the transmission networks. It is rare to find them in the distribution networks, since there has never been a real need for them to be placed there. Recently, PMUs can be found in the distribution network, but not in a quantity or density to provide all the information required by future demands. Furthermore, smart meters are being installed in the majority of developed countries and will improve grid observability. The 15-min data they supply give an insight into power consumption, though the real-time state cannot be inferred. Furthermore, renewables are changing their output on a much shorter time scale. According to [15] empirical measurements are required for several purposes: the diagnosis of specific impacts of DER, establishing a baseline to compare the impacts of a DER validation of new distribution feeder models on ongoing monitoring to support operations and planning the exploration of unknown phenomena on distribution systems. In the transmission grid, the phase angle of the voltage (or current) is the key to the power flow, dynamics, and stability. Voltage phase-angle measurements from the distribution, including the low-voltage network, might address both the well-known and poorly understood problems, such as dynamic instabilities on the distribution grid, to enable new applications in the context of growing distributed intelligence and renewable resource utilization.

Due to the specific characteristics of the distribution system, such as the short power lines between buses, the high R/X ratio, the small voltage differences, etc. (Further discussed in Sect. 2.1), the accuracy of a PMU, normally used in the transmission network, is not sufficient. The resolution of the angle needs to be improved from $0.1°$ to $0.002°$, and the sampling frequency needs to be increased from one sample per cycle to 512 samples per cycle to also capture the system dynamics.

The first step toward controllability of the distribution system is observability, i.e., obtaining accurate knowledge about the relevant parameters of the system. The state

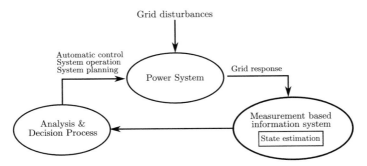

Fig. 1.3 Placement of state estimation in the regulation loop

estimation (SE) solution, which enables full system observability with the required accuracy and resolution, is an essential part of the measurement system. The SE is a signal-processing technique designed to extract the values of a group of parameters (state variables) from a collection of noisy measurements. In the case of power systems, the selected state variables are normally bus-voltage phasors or branch-current phasors. Once the state variables are obtained, it is possible to calculate any quantity of interest and run any control or management function. The SE is expected to play a key role [16] in providing observability for many of the advanced control functions, such as distributed voltage control [17–22], load estimation, fault detection, isolation, and service restoration (FDIR), short-circuit analysis, and many others (see Fig. 1.3). However, this requires increasingly sophisticated measurement instruments and techniques for power-quality monitoring, the rapid detection of anomalous events and, more generally, an accurate network state estimation, especially in a distribution grid [15].

In a densely monitored distribution grid, a wide range of new applications has been identified that could benefit from such accurate measurements in the literature. In Fig. 1.4, we give an overview of the potential applications as seen by the North American Synchro Phasor Initiative (NASPI) [23]. Accurate knowledge about the current situation/status is the baseline for all the actions that need to be taken in order to ensure quality and reliable operation. Still, the economic boundaries of such systems should also be considered. Mounting a PMU device is an expensive procedure in addition to the direct costs related to purchasing such a device. To observe the dynamic circuit behavior the PMU devices need to be densely deployed. Furthermore, the issue of costs related to the transfer of measurements to control centers becomes important [14, 24]. The expected sampling rate for observing dynamic circuit behavior is 512 samples per cycle (one cycle is 1/50 s in Europe and 1/60 s in North America), where many applications require fast, reliable, and secure data transfers, calling for smart grid ICT solutions based on 5G technologies. With the implemented SE, costs related to the number of devices and consequently the number of measurements to transfer can be reduced significantly. Since the initial development of the concept in the early 1970s [25–27], power system SE has become a critical part of the operation and

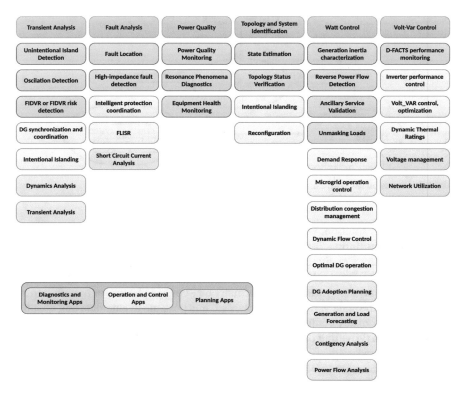

Fig. 1.4 Overview of potential applications

management of transmission systems worldwide. It serves to monitor the state of the grid and enables energy management systems (EMSs) to perform various important control and planning tasks such as establishing near real-time network models for the grid, optimizing power flows, and bad data detection/analysis in the transmission grids. In our vision of future grids the distribution system's state estimation is one of the core elements. With high sampling rates, the state estimation can even be used for some dynamic circuit-behavior analysis-based applications.

1.2 Motivation and Goals

National economies and people's lifestyles require electricity to be accessible, reliable, and affordable. Due to the increased penetration of distributed energy resources and flexible load, power-distribution systems have grown in complexity. Thus, to manage distribution systems in real time, operators need to overcome the challenge of low observability in distribution systems. In addition, it will be necessary to develop a fast, distributed regulation loop that will ensure the safe and stable operation of the system in the future. The first step toward this is to introduce a tool that will make the distribution system observable. One promising solution is the distribution system state estimation. Furthermore, because of the amount of data coming from smart meters, distributed generation measurements, switches, etc., the ideal distribution state estimation methods need to be able to process heterogeneous data. We believe that it is possible to develop a state estimation tool and to design a measurement configuration in such way that their combination will overcome the challenges and make the distribution system observable, which in the future will allow the implementation of control functions such as coordinated voltage regulation and load estimation [28].

The purpose of this book is to present solutions to the state-of-the-art problems in state estimation methods for distributed power systems and to evaluate a feasible three-phase distribution system state estimation along with the configuration of measurement devices that would satisfy power systems. Furthermore, the problem of observability in power-distribution networks is investigated. In order to develop, implement, and thoroughly evaluate a three-phase distribution system state estimation (DSSE), the book focuses on the following:

- To present a unified branch model that allows a unified representation of three-phase distribution network elements. To present conductors, transformers, tap changers, and voltage regulators in any connection, with the same model, from which the branch-admittance matrices can then be derived. To represent voltage regulators and tap changers as a single branch and, in comparison with existing approaches, eliminate the need to introduce an additional bus.
- To develop a three-phase DSSE model that works with heterogeneous measurement types (active and reactive power injections, branch currents, active and reactive power flows, and bus voltages) and is suitable to handle large distribution networks in near real time.
- To evaluate the impact of model and measurement uncertainties for different measurement configurations and different state estimation algorithms. To implement a method that evaluates all of the sensitivities at once, and is feasible for the evaluation of large distribution networks. We also provide guidelines for the design of measurement configurations that can reduce the impact of the uncertainties of the model and measurements in the three-phase DSSE. In addition, an upper error bound in interval sensitivity analysis is suggested.

The results presented in this book should contribute to further research and the development of methods, models, procedures, and algorithms for SE in distribution networks. It is also worth noting that the proposed methods were implemented in a real network within the FP7 project SUNSEED (Sustainable and robust networking for smart electricity distribution) [14, 24], which proposed an evolutionary approach to the utilization of the already present communication networks from both energy and telecom operators. The SE system was applied to a part of the distribution grid. The test bed contained 31 buses, 11 distribution transformers, 5 solar plants, and co-generation. The results were promising, however, an in-depth analysis of the use case is beyond the scope of this book.

1.3 Book Summary by Chapters

The topics are collected in five chapters that are organized according to the following outline. In this chapter, we present the evolution of power-distribution systems from unidirectional to bidirectional power flows, which are characterized by distributed energy resources and flexible loads. Furthermore, the motivation and goals for the development of a distribution system's state estimation are described together with the potential applications, which can be implemented in a densely monitored distribution grid.

In Chap. 2 we position the state estimation (SE) in the context of signal processing, and its relation to power systems. The main differences between the transmission and distribution networks are highlighted, and the main challenges of the introduction of SE into distribution systems are highlighted. Different types of estimators are reviewed, i.e., MLE, WLS, LAV, and SHGM, based on projection statistics. Besides their statistical efficiency, the used estimators are also classified in terms of their robustness. In addition, models for three-phase underground cables, overhead lines, transformers, tap changers, and voltage regulators are reviewed. A unified, three-phase branch model as a generalization of the existing three-phase line models is presented. It enables the modeling of voltage regulators or tap changers and three-phase conductors or transformers on the same branch without the introduction of an extra support bus.

In Chap. 3, the numerical solution algorithms for the reviewed estimators are presented. Different numerical algorithms are compared, and considerations with respect to condition numbers, measurement configurations, speed of computation, and statistical efficiency are given. In addition, the implementation and performance evaluation of different state estimation methods on the proposed unified three-phase branch model is presented. In order to apply the numerical solution to our problem, first, the partial derivatives with respect to the state variables were calculated. Numerical simulations were performed for all the estimators, with different measurement configurations on the IEEE test feeder networks having 13 and 123 buses.

Chapter 4 investigates the impact of model and measurement uncertainties on the state estimation results for different estimators and different measurement configurations in a three-phase distribution network. To obtain all the sensitivities of interest, the approach with a perturbation of the Karush–Kuhn–Tucker conditions is implemented. The sensitivities obtained are expressed in terms of uncertainty intervals rather than just partial derivatives. Based on the results and selected optimization criterion the best estimator is suggested and an upper error bound is proposed on the interval analysis.

Chapter 5 provides final comments and considerations based on the results and concludes the book.

References

1. Engineering National Academies of Sciences and Medicine, *Enhancing the Resilience of the Nations Electricity System* (The National Academies Press, Washington, DC, 2017). ISBN: 978-0-309-46307-2. https://doi.org/10.17226/24836
2. D. Friedman, 4 Charts That Show Renewable Energy is on the Rise in America (2018), https://energy.gov/eere/articles/4-charts-show-renewable-energy-rise-america. Accessed 01 Sept 2018
3. International Energy Agency (IEA), Global EV Outlook (2018), https://www.iea.org/publications/freepublications/publication/Global_EV_Outlook_2016.pdf. Accessed 01 Sept 2018
4. G. Haddadian, M. Khodayar, M. Shahidehpour, Accelerating the global adoption of electric vehicles: barriers and drivers. Electr. J. **28**(10), 53–68 (2015). ISSN: 10406190. https://doi.org/10.1016/j.tej.2015.11.011
5. L. Pieltain Fernandez et al., Assessment of the impact of plug-in electric vehicles on distribution networks. IEEE Trans. Power Syst. **26**(1), 206–213 (2011). ISSN: 0885-8950, 1558-0679. https://doi.org/10.1109/TPWRS.2010.2049133. Accessed 14 June 2017
6. C.L. Masters, Voltage rise: the big issue when connecting embedded generation to long 11 kV overhead lines. Power Eng. J. **16**(1), 5–12 (2002). Accessed 17 May 2015
7. M.H.J. Bollen, N. Etherden, Overload and overvoltage in low-voltage and medium-voltage networks due to renewable energy - some illustrative case studies, in *2011 2nd IEEE PES International Conference and Exhibition on Innovative Smart Grid Technologies (ISGT Europe)* (IEEE, 2011), pp. 1–8. Accessed 16 Nov 2015
8. N. Kreifels et al., Analysis of photovoltaics and wind power in future renewable energy scenarios. Energy Technol. **2**(1), 29–33 (2014). ISSN: 21944288. https://doi.org/10.1002/ente.201300090. Accessed 15 June 2017
9. A. Oudalov, D. Chartouni, C. Ohler, Optimizing a battery energy storage system for primary frequency control. IEEE Trans. Power Syst. **22**(3), 1259–1266 (2007). ISSN: 0885- 8950. https://doi.org/10.1109/TPWRS.2007.901459. Accessed 06 Oct 2016
10. H.-J. Kunisch, K.G. Kramer, H. Dominik, Battery energy storage another option for load-frequency-control and instantaneous reserve. IEEE Trans. Energy Convers. **3**, 41–46 (1986). Accessed 06 Oct 2016
11. T. Borsche et al., Power and energy capacity requirements of storages providing frequency control reserves, in *IEEE PES General Meeting* (Citeseer, Vancouver, 2013). Accessed 06 Oct 2016
12. Y. Mu et al., Primary frequency response from electric vehicles in the great britain power system. IEEE Trans. Smart Grid **4**(2), 1142–1150 (2013). ISSN: 1949-3053, 1949-3061. https://doi.org/10.1109/TSG.2012.2220867
13. M. Pau, State estimation in electrical distribution systems. Ph.D. Thesis. University of Cagliari, Cagliari, Italy, 2015
14. J.J. Nielsen et al., Secure real-time monitoring and management of smart distribution grid using shared cellular networks. IEEE Wirel. Commun. **24**(2), 10–17 (2017). ISSN: 1536-1284. https://doi.org/10.1109/MWC.2017.1600252
15. A. Von Meier et al., Micro-synchrophasors for distribution systems. English, in *2014 IEEE PES Innovative Smart Grid Technologies Conference, ISGT 2014* (Cited By: 147) (2014)
16. J. Fan, S. Borlase, The evolution of distribution. IEEE Power Energy Mag. **7**(2), 63–68 (2009). ISSN: 1540-7977. https://doi.org/10.1109/MPE.2008.931392. Accessed 20 May 2015
17. A. Kulmala et al., Demonstrating coordinated voltage control in a real distribution network, in *2012 3rd IEEE PES International Conference and Exhibition on Innovative Smart Grid Technologies (ISGT Europe)* (IEEE, 2012), pp. 1–8. Accessed 26 Nov 2015
18. R. Caldon et al., Co-ordinated voltage regulation in distribution networks with embedded generation, in *18th International Conference and Exhibition on Electricity Distribution, CIRED 2005* (IET, 2005), pp. 1–4. Accessed 26 Nov 2015

19. C. Gao, M.A. Redfern, A review of voltage control techniques of networks with distributed generations using On-Load Tap Changer transformers, in *2010 45th International Universities Power Engineering Conference (UPEC)* (IEEE, 2010), pp. 1–6. Accessed 26 Nov 2015
20. M. Meuser, H. Vennegeerts, P. Schafer, Impact of voltage control by distributed generation on hosting capacitiy and reactive power balance in distribution grids, in *Integration of Renewables into the Distribution Grid, CIRED 2012 Workshop* (IET, 2012), pp. 1–5. Accessed 16 Nov 2015
21. A. Kulmala, S. Repo, P. Jarventausta, Active voltage control - from theory to practice (2012) Accessed 16 Nov 2015
22. A. Kulmala et al., Active voltage level management of distribution networks with distributed generation using on load tap changing transformers, in *Power Tech, 2007 IEEE Lausanne* (IEEE, 2007), pp. 455–460. Accessed 26 Nov 2015
23. Website, North American SynchroPhasor Initiative (NASPI) (2019), https://www.naspi.org/. Accessed 25 Nov 2019
24. Website, Fp7 SUNSEED (2019), http://sunseed-fp7.eu/. Accessed 25 Nov 2019
25. F.C. Schweppe, Power system static-state estimation part I: the exact model. IEEE Trans. Power Appar. Syst. ()
26. F.C. Schweppe, B. Rom Douglas, Power system static-state estimation part II: the approx. model. IEEE Trans. Power Appar. Syst. ()
27. F.C. Schweppe, Power system static-state estimation, part iii: implementation. IEEE Trans. Power Appar. Syst. **1**, 130–135 (1970). Accessed 06 Nov 2014
28. U. Kuhar, Three-phase state estimation in power distribution systems. Ph.D. Thesis. Jožef Stefan International Postgraduate School, Ljubljana, Slovenia, 2018

Chapter 2
State Estimation

2.1 Challenges of State Estimation in Distribution Systems

Estimation theory can be found at the heart of many signal-processing systems that are designed to extract the values of a group of parameters from noisy measurements. Statistical state estimation (SE) was first applied to power systems back in 1970 by Schweppe and others. It was proposed for the real-time monitoring of electric power-transmission networks [1–3]. The proposed approach quickly gained traction and has since remained an extremely active area of research. The so-called static single-phase weighted least squares (WLSs) estimation of power-transmission networks is now the standard in the electric power industry. It is adopted by virtually every transmission system operator (TSO).

SE was identified as a tool that can detect and reduce the errors in measurement devices and topology models. Measurements are always subject to errors due to the limited accuracy, aging, and malfunctioning of devices. Moreover, communication errors can occur, and in cases where information about a specific bus is obtained from a single measurement device, the network operator loses all the information about that bus when, for whatever reason, the measurements become unavailable.

Although the term SE in general refers to an algorithm or mathematical expression that extracts some predetermined parameters of interest from noisy measurement data, in terms of power systems, SE represents a comprehensive system with a number of additional functionalities besides the core estimation algorithm. These functionalities may vary from system to system, but they most commonly include the following [4]:

- *Topology processor*: builds a network topology (i.e., a system-admittance matrix) based on the latest information on the states of the switches, circuit breakers, transformer taps, and regulator settings.
- *Observability analysis*: is a pre-calculation step that makes sure the system is observable, i.e., there are enough real-time and pseudo-measurements that the SE algorithm will converge on a solution.

© The Author(s), under exclusive license to Springer Nature Switzerland AG 2020
U. Kuhar et al., *Observability of Power-Distribution Systems*,
SpringerBriefs in Applied Sciences and Technology,
https://doi.org/10.1007/978-3-030-39476-9_2

- *SE algorithm*: is the core of the SE system. It takes all the provided data (real-time measurements, pseudo-measurements, and topology) and estimates the most probable state of the system in terms of predetermined criteria.
- *Detection of bad data and topology errors*: is a pre-calculation or post-calculation step that determines whether a measurement or a network component's status (e.g., circuit breaker) has a value that is not likely to be correct or in the expected range.
- *Database with pseudo-measurements*: holds the pseudo-measurement values relevant to specific points in time. This data is normally obtained by means other than just measurement devices. As an example, historical measurements are fed into the machine-learning system, which can determine the most probable current values, taking into account the calendar data, weather data, etc.
- *Simulation and Contingency Analysis*: is a post-calculation step that enables the planning of contingency scenarios given the current trends in a power system. This enables the operator to anticipate problems and act before they occur.

Figure 2.1 depicts the relationships for the listed functionalities, and the propagation of data through them.

Conceptually, SE has the same function in both transmission and distribution systems, but due to some inherent differences between the two, SE for transmission systems cannot be straightforwardly applied to distribution systems. In the case of SE application, the most relevant differences between the two systems are [5].

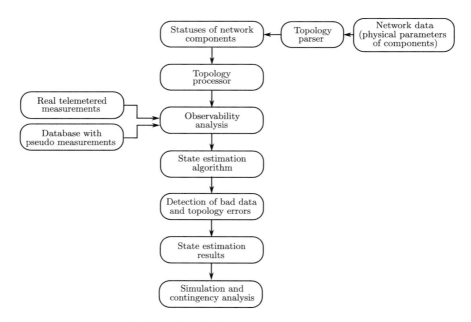

Fig. 2.1 Main components of a state estimation system

- *Imbalance*: A power grid is a three-phase network. In transmission grids, the quantities in phases are assumed to be balanced, because of the averaging effect of many consumers and transformer connections, which can balance their loading to some extent. Transmission networks are modeled through an equivalent positive sequence phasor in symmetrical components. In contrast, distribution grids extend directly to consumers, thus, single-phase loads and unsymmetrical loading are reflected much more in the network. In order to obtain an accurate state of the system, the model needs to account for possible asymmetries and, thus, all three phases need to be modeled. Most of the approaches in the literature [6–10] use the so-called equivalent-measurements approach, where all the measurements are converted into current measurements. This approach, however, requires an extra computational step. Thus, in the book we propose a model for the unified representation of the elements of a three-phase distribution network [11]. Conductors, transformers, tap changers, and voltage regulators, in any connection, are all presented with the same model, from which the branch-admittance matrices are derived. With the proposed model, voltage regulators or tap changers and conductor sections or transformers connected in series can be represented as a single branch.
- *Branch parameters*: An electric line has four parameters that affect its ability to fulfill its function as part of a power system, i.e., series resistance and inductance, and shunt capacitance and conductance [12]. Since the power flows and voltage levels of distribution and transmission networks are different, the conductors have different properties. In transmission networks the series impedance has a prevailing inductance term. In terms of the ratio of reactance (X) to resistance (R), i.e., (X/R), the value is approximately 10 in transmission networks. This allows for the so-called decoupled estimation, where the estimation of the voltage magnitudes and angles is separated [13]. Due to the relatively high reactance term, the phase-angle spread is also greater along the line. This makes phase-angle differences easier to measure. In distribution networks, the X/R ratios are significantly lower (of the order of $0.1-1$). As a result, the branch model cannot be simplified to allow for a fast decoupled estimation. Also, since the phase-angle spread along the line is much smaller, in the case of phase-angle measurements, there is a need for very accurate measurements.
- *High number of nodes*: Distribution systems commonly have ratios of the order of $10^4 - 10^6$ buses, as opposed to transmission networks that normally have ratios of the order of 100 buses. This number is further multiplied by ≈ 3 since all three phases need to be modeled in the distribution network. This characteristic poses two separate problems. The problem's dimension corresponds to the number of buses. Most solution techniques have a computational complexity of $O(n^3)$. The characteristic that can be exploited in addressing this problem is that distribution networks mostly have a radial topology. The matrices are, thus, very sparse and the data dependency is low. This offers a great opportunity for the parallelization of the solution algorithm. The other challenge to overcome is the small number of measurement devices. Distribution networks are currently poorly monitored. There is little presence of measurement instruments in medium- and low-voltage

networks. In the European Union (EU) there is an effort underway to equip all households with advanced measurement infrastructure (AMI) smart meters (SM). For measurements deeper in the network, the crucial factor is price. Because of the large number of nodes, the redundancy of measurements is hard to achieve. One technique to mitigate this issue is so-called pseudo-measurements. These are based on historical data and mathematical models [14, 15]. However, the accuracy of this data is orders of magnitude worse than that of real measurements.

- *Network model uncertainty*: network model parameters such as series impedances or shunt admittances (self and mutual) that are derived from topology data (conductor geometry) can be incorrect. This is due to the inaccurate data provided by the manufacturer or the installation crew, inaccurate measurement campaigns, effects of aging, etc. Uncertain model data has an influence over the estimated state variables. Several methods were applied to the problem of sensitivity analysis in a power system's state estimation. In this book, we perform a sensitivity analysis of the state estimation results to small deviations from model or measurement assumptions. Deviations up to a few tens of percent are considered as small. An example of such a deviation would be the incorrectly modeled length of a conductor.

2.2   Problem Statement

Mathematically, the state estimation problem can be stated by extracting an unknown vector of system states $\mathbf{x} = [x_1, x_2, \ldots, x_n]$ from the m-point dataset $\{f_1, f_2, \ldots, f_m\}$ that depends on a vector. Thus, to determine $\mathbf{x}$ based on the data, an estimator is defined as

$$\hat{\mathbf{x}} = \mathbf{g}(\{f_1, f_2, \ldots, f_n\}) \tag{2.1}$$

where $\mathbf{g}$ is a function vector that needs to be determined. In the process of estimator determination, the first step is to mathematically model the data. Since the data are inherently random, they are modeled by the probability density function (PDF). The PDF is parameterized by the unknown vector of parameters $\mathbf{x}$. As an example, if there is only one data point and x denotes the mean, then the PDF of the data can be

$$p(f_1; x) = \frac{1}{\sqrt{2\pi\sigma^2}} exp\left[-\frac{1}{2\sigma^2}(f_1 - x)^2\right]. \tag{2.2}$$

Since the value of x affects the probability of f_1, it is possible to infer the value of x from the observed value of f_1. This specification of the PDF is critical in determining an estimator that will be optimal in a certain sense. An estimation based on PDFs is termed a *classical* estimation in which the parameters of interest are assumed to be *deterministic* but unknown [16].

There are different classes of estimators that can be derived for a particular estimation problem. In the case of the classical approach, Fig. 2.2 depicts the decision

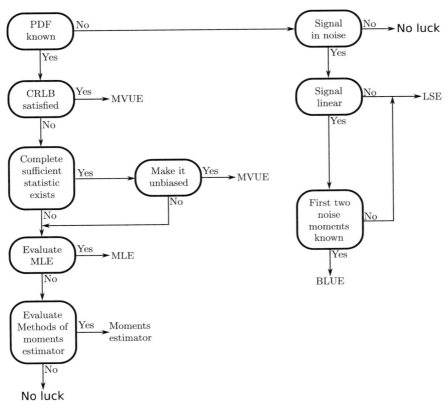

Fig. 2.2   Selection of the estimator

flowchart, including different estimators, and a decision logic that leads to their selection. In our case, we will first consider the case where we assume that the measurement errors have a Gaussian distribution. Considering the flowchart for the selection of the estimator class, we will select the maximum-likelihood estimation (MLE). Further, it will be shown that if the assumption about the PDF of the measurement errors is dropped, the estimator obtained from the MLE is actually the weighted least squares (WLS) estimator.

Besides the optimality criterion, another important aspect of the state estimation algorithm is its robustness. Robustness is defined as the insensitivity of the state estimates to major deviations in a limited number of redundant measurements [17]. In this chapter, two estimators from the class of robust estimators, i.e., the least absolute value (LAV) and the Schweppe Huber Generalized-M (SHGM), and their solution methods are reviewed. Robustness concepts, such as outliers and leverage points, and methods for detecting them are discussed in more detail in [17].

2.2.1 Data Model

In this subsection a data model for state estimation in distribution power networks will be defined.

The network state can be inferred from a set of measurements $(f_1, f_2, \ldots, f_m)$ that have an established (nonlinear in general) relation to the selected state variables $(x_1, x_2, \ldots, x_n)$:

$$
\mathbf{f} = \begin{bmatrix} f_1 \\ f_2 \\ \vdots \\ f_m \end{bmatrix} = \begin{bmatrix} h_1(x_1, \ldots, x_n) \\ h_2(x_1, \ldots, x_n) \\ \vdots \\ h_m(x_1, \ldots, x_n) \end{bmatrix} + \begin{bmatrix} w_1 \\ w_2 \\ \vdots \\ w_m \end{bmatrix} \tag{2.3}
$$
$$
= \mathbf{h}(\mathbf{x}) + \mathbf{w}
$$

The symbol $\mathbf{w} \sim \mathcal{N}(\mathbf{0}, \mathbf{R})$ represents the noise contribution to the measurement vector. It is assumed that the noise in the measurements is uncorrelated and, thus, the measurement covariance matrix is $\mathbf{R} = diag\{\sigma_1^2, \sigma_2^2, \ldots, \sigma_m^2\}$. In Sect. 2.3, measurement functions $h(x_1, \ldots, x_n)$ that relate different measurement types with the selected state variables will be developed.

2.2.2 Maximum-Likelihood Estimation

The MLE is defined as the value of x (or $\mathbf{x}$) that maximizes the likelihood function. The likelihood function is specified as a PDF. When a closed-form expression cannot be found for the MLE, a numerical approach can be utilized for an iterative maximization of the likelihood function.

For the data model defined in Sect. 2.2.1, a likelihood function can be written as

$$
p(\mathbf{f}; \mathbf{x}) = \prod_{i=1}^{m} \frac{1}{\sqrt{2\pi\sigma_i^2}} exp\left[-\frac{1}{2\sigma_i^2}(f_i - h_i(\mathbf{x}))^2 \right]
$$
$$
= \prod_{i=1}^{m} \frac{1}{\sqrt{2\pi\sigma_i^2}} exp\left[-\frac{1}{2}(\mathbf{f} - \mathbf{h}(\mathbf{x}))^T \mathbf{R}(\mathbf{f} - \mathbf{h}(\mathbf{x})) \right]. \tag{2.4}
$$

assuming the differentiability of the likelihood function, the MLE is found from

$$
\frac{\partial \ln p(\mathbf{f}; \mathbf{x})}{\partial \mathbf{x}} = \mathbf{0}. \tag{2.5}
$$

Considering Eq. 2.5, first the logarithm is taken from Eq. 2.4

$$\ln p(\mathbf{f}; \mathbf{x}) = \sum_{i=1}^{m} \frac{1}{\sqrt{2\pi \sigma_i^2}} + \left[-\frac{1}{2} (\mathbf{f} - \mathbf{h}(\mathbf{x}))^T \mathbf{R}(\mathbf{f} - \mathbf{h}(\mathbf{x})) \right]. \tag{2.6}$$

We define a new function $J_{MLE}(\mathbf{x})$ as

$$J_{MLE}(\mathbf{x}) = \left[-\frac{1}{2}(\mathbf{f} - \mathbf{h}(\mathbf{x}))^T \mathbf{R}(\mathbf{f} - \mathbf{h}(\mathbf{x})) \right]. \tag{2.7}$$

When the derivative of Eq. 2.4 with respect to $\mathbf{x}$ is taken, Eq. 2.5 equals

$$\frac{\partial \ln p(\mathbf{f}; \mathbf{x})}{\partial \mathbf{x}} = \frac{\partial J(\mathbf{x})}{\partial \mathbf{x}} = \mathbf{0}. \tag{2.8}$$

In the case of a linear measurement model (where $\mathbf{h}(\mathbf{x}) = \mathbf{Hx}$), a closed-form solution can be found to Eq. 2.5. But in a more general case, where the measurement model is nonlinear, we need to resort to numerical methods, such as Newton–Raphson or Gauss–Newton.

2.2.3 Weighted Least Squares Estimation

The method of least squares (LS) or WLS makes no probabilistic assumptions about the data. The only thing that is assumed is a signal model. The disadvantage of this is that no claims about optimality can be made and the statistical performance cannot be assessed. However, the method is easy to implement and is, thus, widely used in practice.

The LS or WLS approach only attempts to minimize the squared difference between the given data $\mathbf{f} = (f_1, f_2, \ldots, f_m)$ and the assumed signal model. The signal is deterministic and depends on the unknown vector of parameters $\mathbf{x}$. The LS estimator chooses the $\mathbf{x}$ that makes the signal model $\mathbf{h}(\mathbf{x})$ closest to the observed data $\mathbf{f}$. The closeness is measured with the LS error criterion:

$$J_{LS}(\mathbf{x}) = \sum_{i=1}^{m} (f_i - h_i(\mathbf{x}))^2 = (\mathbf{f} - \mathbf{h}(\mathbf{x}))^T (\mathbf{f} - \mathbf{h}(\mathbf{x})) \tag{2.9}$$

In the case where we have different levels of confidence in the measurements, the WLS criterion is used

$$J_{WLS}(\mathbf{x}) = \sum_{i=1}^{m} r_i^{-1} (f_i - h_i(\mathbf{x}))^2 = (\mathbf{f} - \mathbf{h}(\mathbf{x}))^T \mathbf{R}^{-1} (\mathbf{f} - \mathbf{h}(\mathbf{x})). \tag{2.10}$$

Note that if $\mathbf{f} - \mathbf{h}(\mathbf{x}) \sim \mathcal{N}(\mathbf{0}, \sigma^2\mathbf{I})$, Eq. 2.7 is practically identical to Eq. 2.10 and, thus, the WLS estimator is also the MLE. Obviously, the same solution methods apply for the minimization of Eq. 2.10 (or Eq. 2.9) as for the solution of Eq. 2.8.

2.2.4   Least Absolute Value Estimation

Like the LS (WLS) estimator, the Least Absolute Value (LAV) estimator makes no probabilistic assumptions about the data, and the only thing that is assumed is the signal model. The LAV approach attempts to minimize the absolute difference between the given data $\mathbf{f} = (f_1, f_2, \ldots, f_m)$ and the assumed signal model. The signal is deterministic and depends on the unknown vector of parameters $\mathbf{x}$. The LAV estimator chooses the $\mathbf{x}$ that makes the signal model $\mathbf{h}(\mathbf{x})$ closest to the observed data $\mathbf{f}$. The closeness is measured with the LAV criterion:

$$J_{LAV}(\mathbf{x}) = \sum_{i=1}^{m} \left| f_i - h_i(\mathbf{x}) \right|. \tag{2.11}$$

The value of the LAV estimator stems from its robustness in the face of outliers. It reduces the influence of bad data on the state estimation results and belongs to the class of robust estimators. In terms of solution, this problem can be recast as a linear programming problem that can be solved with any of the existing techniques, such as the primal-dual interior-point method, the iteratively reweighted least squares (IRLS), or a simplex-based algorithm. In Sect. 3.5, an IRLS method for solving the LAV estimator is reviewed.

2.2.5   Schweppe–Huber's Generalized-M Estimation Based on Projection Statistics

The Schweppe–Huber's Generalized-M (SHGM) estimator is another from the class of robust estimators. It implements a projection algorithm that accounts for the sparsity of the Jacobian matrix. It assigns to each data point a projection statistic. Based on these projection statistics, a robustly weighted estimator is defined. This estimator does not downweight good leverage points while bounding the influence of the bad ones [18].

The SHGM estimator attempts to minimize the following objective function:

$$J_{SHGM}(\mathbf{x}, \mathbf{a}) = \sum_{i=1}^{m} \gamma_i^2 \rho(r_i) \tag{2.12}$$

where the residual is defined as

$$r_i = \frac{f_i - h_i(\mathbf{x})}{\gamma_i},$$ (2.13)

and the ρ function is defined as

$$\rho(r_i) = \begin{cases} \frac{1}{2}r_i^2 & |r_i| \le c \\ c|r_i| - \frac{c^2}{2} & \text{otherwise} \end{cases}$$ (2.14)

where

$$\gamma_i = \min\left\{1, \left(\frac{b_i}{PS_i}\right)^2\right\},$$ (2.15)

$$b_i = \chi_{\xi,0.975}^2,$$ (2.16)

$$PS_i = \max_{\mathbf{H}_k} \frac{|\mathbf{H}_i \mathbf{H}_k^T|}{\beta_i} \quad k = 1, 2, \ldots, m,$$ (2.17)

$$\beta_i = 1.1926 \cdot \text{lomed}_i \left\{\text{lomed}_{i \ne j}|\mathbf{H}_i\mathbf{H}_k^T + \mathbf{H}_j\mathbf{H}_k^T|\right\},$$ (2.18)

$\mathbf{H}_i$ is the ith row vector of the Jacobian matrix $\mathbf{H}$, ξ is the number of nonzero entries in the corresponding (ith) row of the Jacobian matrix, the lomed function is the low median of the given set, and a good choice for c ranges from 1 to 3 [18].

2.3 Network Model

In this section, a multiphase model of the distribution network will be developed. First, a unified three-phase branch model will be presented; then, model parameters for all the relevant network components, such as conductors, transformers, voltage regulators, and loads, will be developed. Further, its use to build a complete network model will be demonstrated.

2.3.1 A Unified Three-Phase Branch Model

A unified three-phase branch model is proposed here. The model can represent all the elements of a three-phase distribution network. Conductors, transformers, tap changers, and voltage regulators in any connection, are all presented with the same model from which the branch-admittance matrices are derived. With the proposed model, voltage regulators or tap changers and conductor sections or transformers connected in series can be represented as a single branch (as is usually found in practice and also in IEEE distribution test feeders). In comparison with the existing approaches, we eliminate the requirement to introduce an additional bus (where a voltage regulator or a tap changer, and a conductor or a transformer are modeled as two separate branches), which increases the dimension of the problem. This is also beneficial from the computer-implementation point of view, since it simplifies the computer program's design.

Three-phase branch models that have been developed for the power-flow analysis of distribution networks using a forward/backward sweep algorithm [19] are not applicable for this purpose in a straightforward way, because matrix transformations are needed when singularities are encountered for several transformer models [20]. A modified version of the forward/backward sweep algorithm can be used for state estimation problems, but it lacks accuracy, as shown in [21]. The references [22], [23] claim the use of three-phase component models with a WLS approach or a variation, but the relationships between the measurements and the state variables are not provided.

The proposed three-phase branch model is shown in Fig. 2.3. The model is symmetrical and establishes a relationship between the voltages and the currents on the nodes k and m. To simplify the formal derivation, the auxiliary p and q nodes are introduced. The proposed model is a three-phase extension of the single-phase, unified, branch model depicted in Fig. 2.4 that was developed by Monticelli in [24], and can be applied to networks where the phase symmetry is assumed. Here, line-to-ground voltages are selected as the state variables and, thus, the derivations of other variables depend on the voltage phasors. The bus-voltage phasors are denoted as

$$E_m^a = |V_m^a| \cdot e^{j\phi_m^a} \tag{2.19}$$

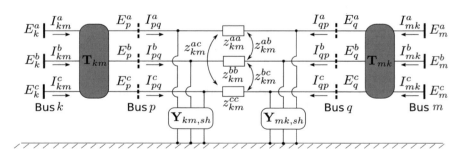

Fig. 2.3 Three-phase unified branch model

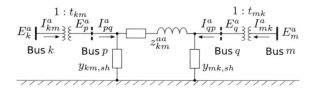

Fig. 2.4 Single-phase unified branch model

where $|V_m^a|$ represents the magnitude of the phasor, and ϕ_m^a represents a phase angle at the node m on phase a. The elements $\mathbf{T}_{km}$ and $\mathbf{T}_{mk}$ are presented as matrices (Eq. 2.20) and are used for the modeling of voltage regulators or tap changers. Each phase voltage or current on the auxiliary node can be a linear combination of all three voltages or currents of the boundary node on their respective sides. This is necessary in order to correctly model the different regulator connections and tap settings.

$$\mathbf{T}_{km} = \begin{bmatrix} t_{km}^{aa} & t_{km}^{ab} & t_{km}^{ac} \\ t_{km}^{ba} & t_{km}^{bb} & t_{km}^{bc} \\ t_{km}^{ca} & t_{km}^{cb} & t_{km}^{cc} \end{bmatrix} \quad \mathbf{T}_{mk} = \begin{bmatrix} t_{mk}^{aa} & t_{mk}^{ab} & t_{mk}^{ac} \\ t_{mk}^{ba} & t_{mk}^{bb} & t_{mk}^{bc} \\ t_{mk}^{ca} & t_{mk}^{cb} & t_{mk}^{cc} \end{bmatrix} \tag{2.20}$$

The series impedance $\mathbf{Z}_{km}$ and the shunt-admittance $\mathbf{Y}_{km,sh}$, $\mathbf{Y}_{mk,sh}$ matrices are sized as 3×3 elements:

$$\mathbf{Z}_{km} = \begin{bmatrix} z_{km}^{aa} & z_{km}^{ab} & z_{km}^{ac} \\ z_{km}^{ba} & z_{km}^{bb} & z_{km}^{bc} \\ z_{km}^{ca} & z_{km}^{cb} & z_{km}^{cc} \end{bmatrix} = \mathbf{Y}_{km}^{-1}$$

$$\mathbf{Y}_{km,sh} = \begin{bmatrix} y_{km,sh}^{aa} & y_{km,sh}^{ab} & y_{km,sh}^{ac} \\ y_{km,sh}^{ba} & y_{km,sh}^{bb} & y_{km,sh}^{bc} \\ y_{km,sh}^{ca} & y_{km,sh}^{cb} & y_{km,sh}^{cc} \end{bmatrix} \tag{2.21}$$

$$\mathbf{Y}_{mk,sh} = \begin{bmatrix} y_{mk,sh}^{aa} & y_{mk,sh}^{ab} & y_{mk,sh}^{ac} \\ y_{mk,sh}^{ba} & y_{mk,sh}^{bb} & y_{mk,sh}^{bc} \\ y_{mk,sh}^{ca} & y_{mk,sh}^{cb} & y_{mk,sh}^{cc} \end{bmatrix}$$

2.3.1.1 Current Equations

Since the line-to-ground bus voltages are selected as independent variables, the line currents are derived as functions of the line-to-ground bus voltages. The line currents can be a part of the measurement model in the formulation of the state estimation problem, and are necessary for the derivation of line power flows. The line current for phase a with the support of auxiliary nodes can be written as

$$I_{km}^a = I_{pq}^a \cdot t_{km}^{aa} + I_{pq}^b \cdot t_{km}^{ab} + I_{pq}^c \cdot t_{km}^{ac} = \sum_{j \in \Omega_p} I_{pq}^j \cdot t_{km}^{aj}, \tag{2.22}$$

where $\Omega_p = \{a, b, c\}$ denotes the phases. The current between the auxiliary buses can be expressed as

$$\begin{aligned} I_{pq}^a &= (E_p^a - E_q^a)y_{km}^{aa} + (E_p^b - E_q^b)y_{km}^{ab} + (E_p^c - E_c^a)y_{km}^{ac} \\ &\quad + E_p^a y_{km,sh}^{aa} + E_p^b y_{km,sh}^{ab} + E_p^c y_{km,sh}^{ac} \\ &= \sum_{n \in \Omega_p} \left((E_p^n - E_q^n)y_{km}^{an} + E_p^n y_{km,sh}^{an}\right). \end{aligned} \tag{2.23}$$

The relationship between the voltage phasors at the auxiliary buses and the boundary buses is

$$E_p^a = E_k^a t_{km}^{aa} + E_k^b t_{km}^{ab} + E_k^c t_{km}^{ac} = \sum_{i \in \Omega_p} E_k^i t_{km}^{ai} \tag{2.24}$$

$$E_q^a = E_m^a t_{mk}^{aa} + E_m^b t_{mk}^{ab} + E_m^c t_{mk}^{ac} = \sum_{i \in \Omega_p} E_m^i t_{mk}^{ai}. \tag{2.25}$$

Equations 2.24 and 2.25 are inserted into Eq. 2.23:

$$I_{pq}^a = \sum_{n \in \Omega_p} \left[\left(\sum_{i \in \Omega_p} E_k^i t_{km}^{ni} - \sum_{i \in \Omega_p} E_m^i t_{mk}^{ni} \right) y_{km}^{an} + \sum_{i \in \Omega_p} E_k^i t_{km}^{ni} y_{km,sh}^{an} \right], \tag{2.26}$$

finally, Eq. 2.26 is inserted into Eq. 2.22:

$$I_{km}^a = \sum_{j \in \Omega_p} \sum_{n \in \Omega_p} \sum_{i \in \Omega_p} \left[(y_{km}^{jn} + y_{km,sh}^{jn}) t_{km}^{ni} t_{km}^{aj} E_k^i - t_{mk}^{ni} t_{km}^{aj} y_{km}^{jn} E_m^i \right]. \tag{2.27}$$

The derivation is limited to one phase, as others can be written analogously. Because of the model's symmetry, the I_{mk}^a is simply:

$$I_{mk}^a = \sum_{j \in \Omega_p} \sum_{n \in \Omega_p} \sum_{i \in \Omega_p} \left[(y_{km}^{jn} + y_{mk,sh}^{jn}) t_{mk}^{ni} t_{mk}^{aj} E_m^i - t_{km}^{ni} t_{mk}^{aj} y_{km}^{jn} E_k^i \right]. \tag{2.28}$$

To obtain a branch-admittance matrix, a complete set of equations is written in matrix form in Eq. 2.29, while Eq. 2.30 displays the same set in a more compact form, which is also more convenient for computer implementation.

$$\begin{bmatrix} I_{km}^a \\ I_{km}^b \\ I_{km}^c \\ I_{mk}^a \\ I_{mk}^b \\ I_{mk}^c \end{bmatrix} = \begin{bmatrix} \sum_{j \in \Omega_p} \sum_{n \in \Omega_p} (y_{km}^{jn} + y_{km,sh}^{jn} t_{km}^{na} t_{km}^{aj}) & \cdots & -\sum_{j \in \Omega_p} \sum_{n \in \Omega_p} t_{mk}^{na} t_{km}^{aj} y_{km}^{jn} & \cdots \\ \sum_{j \in \Omega_p} \sum_{n \in \Omega_p} (y_{km}^{jn} + y_{km,sh}^{jn} t_{km}^{na} t_{km}^{bj}) & \cdots & -\sum_{j \in \Omega_p} \sum_{n \in \Omega_p} t_{mk}^{na} t_{km}^{bj} y_{km}^{jn} & \cdots \\ \vdots & \vdots & \vdots & \\ -\sum_{j \in \Omega_p} \sum_{n \in \Omega_p} t_{km}^{na} t_{mk}^{aj} y_{km}^{jn} & \cdots & \sum_{j \in \Omega_p} \sum_{n \in \Omega_p} (y_{km}^{jn} + y_{mk,sh}^{jn} t_{km}^{na} t_{km}^{aj}) & \cdots \\ -\sum_{j \in \Omega_p} \sum_{n \in \Omega_p} t_{km}^{na} t_{mk}^{bj} y_{km}^{jn} & \cdots & \sum_{j \in \Omega_p} \sum_{n \in \Omega_p} (y_{km}^{jn} + y_{mk,sh}^{jn} t_{km}^{na} t_{km}^{bj}) & \cdots \\ \vdots & \vdots & \vdots & \end{bmatrix} \cdot \begin{bmatrix} E_k^a \\ E_k^b \\ E_k^c \\ E_m^a \\ E_m^b \\ E_m^c \end{bmatrix}$$
$$\tag{2.29}$$

$$\begin{bmatrix} \mathbf{I}_{km}^{\{a,b,c\}} \\ \mathbf{I}_{mk}^{\{a,b,c\}} \end{bmatrix} = \underbrace{\begin{bmatrix} \mathbf{T}_{km} \cdot (\mathbf{Y}_{km} + \mathbf{Y}_{km,sh}) \cdot \mathbf{T}_{km} & -\mathbf{T}_{km} \cdot \mathbf{Y}_{km} \cdot \mathbf{T}_{mk} \\ -\mathbf{T}_{mk} \cdot \mathbf{Y}_{km} \cdot \mathbf{T}_{km} & \mathbf{T}_{mk} \cdot (\mathbf{Y}_{km} + \mathbf{Y}_{mk,sh}) \cdot \mathbf{T}_{mk} \end{bmatrix}}_{\mathbf{Y}_{b,km}} \begin{bmatrix} \mathbf{E}_k^{\{a,b,c\}} \\ \mathbf{E}_m^{\{a,b,c\}} \end{bmatrix}$$
$$\tag{2.30}$$

Another quantity that is often required by applications that use the model is the line power flow. A relationship between the state variables and the active and reactive

branch power flows is derived as follows. The conjugate of the complex power flow
can be written as

$$(S_{km}^a)^* = (E_k^a)^* \cdot I_{km}^a = P_{km}^a - jQ_{km}^a. \tag{2.31}$$

Considering Eq. 2.19 and $y_{km}^{aa} = g_{km}^{aa} + jb_{km}^{aa}$, Eq. 2.31 yields an expression for the real
(Eq. 2.32) and reactive (Eq. 2.33) power flows

$$
\begin{aligned}
P_{km}^a = |V_k^a| \sum_{j \in \Omega_p} \sum_{n \in \Omega_p} \sum_{i \in \Omega_p} \Big\{ & t_{km}^{ni} t_{km}^{aj} |V_k^i| \Big[(g_{km}^{jn} + g_{km,sh}^{jn}) \cos(\phi_k^a - \phi_k^i) \\
& + (b_{km}^{jn} + b_{km,sh}^{jn}) \sin(\phi_k^a - \phi_k^i) \Big] \\
& - t_{mk}^{ni} t_{km}^{aj} |V_m^i| \Big[g_{km}^{jn} \cos(\phi_k^a - \phi_m^i) + b_{km}^{jn} \sin(\phi_k^a - \phi_m^i) \Big] \Big\},
\end{aligned}
\tag{2.32}
$$

$$
\begin{aligned}
Q_{km}^a = |V_k^a| \sum_{j \in \Omega_p} \sum_{n \in \Omega_p} \sum_{i \in \Omega_p} \Big\{ & t_{km}^{ni} t_{km}^{aj} |V_k^i| \Big[(g_{km}^{jn} + g_{km,sh}^{jn}) \sin(\phi_k^a - \phi_k^i) \\
& - (b_{km}^{jn} + b_{km,sh}^{jn}) \cos(\phi_k^a - \phi_k^i) \Big] \\
& - t_{mk}^{ni} t_{km}^{aj} |V_m^i| \Big[g_{km}^{jn} \sin(\phi_k^a - \phi_m^i) - b_{km}^{jn} \cos(\phi_k^a - \phi_m^i) \Big] \Big\}.
\end{aligned}
\tag{2.33}
$$

2.3.2   Conductors

The modeling of distribution overhead lines and underground cables is a critical step
that must be taken before any analysis of the distribution feeder can take place. Here,
the process of determining the series-impedance and shunt-admittance matrices will
be reviewed. Further, geometrical configurations of overhead lines and underground
cables in IEEE reference distribution feeders will be reviewed. The series-impedance
and shunt-admittance matrices for the unified branch model will be developed for
those configurations.

2.3.2.1   Modified Carson's Equations

Carson's equations can be used to determine the self and mutual impedances for
an arbitrary number of overhead or underground conductors. Here, only the final
equations will be reviewed, and the derivation can be found in [19].

Self and mutual series-impedance matrix elements are determined as

$$\hat{z}_{ii-usa} = r_i + 0.0953016 + j0.1213422 \left(\ln \frac{1}{GMR_i} + 7.93401 \right) \Omega/\text{mile}, \tag{2.34}$$

$$\hat{z}_{ii-eu} = r_i + 0.049348 + j0.062832\left(\ln\frac{1}{GMR_i} + 6.8371\right)\Omega/\text{km}, \qquad (2.35)$$

$$\hat{z}_{ij-usa} = 0.0953016 + j0.1213422\left(\ln\frac{1}{GMR_i} + 7.93401\right)\Omega/\text{mile}, \qquad (2.36)$$

$$\hat{z}_{ij-eu} = 0.049348 + j0.062832\left(\ln\frac{1}{D_{ij}} + 6.8371\right)\Omega/\text{km}, \qquad (2.37)$$

where

- GMR_i ...is the geometric mean radius of conductor i in feet in Eq. 2.34, and in meters in Eq. 2.35
- D_{ij} ...is the distance between the conductors i and j in feet in Eq. 2.36, and in meters in Eq. 2.37

2.3.2.2 Series-Impedance Matrices of Overhead Lines in IEEE Distribution Test Feeders

The process of developing the series-impedance matrix includes the following steps. First, conductor types and their geometric layout need to be determined. Then, primitive matrix elements are determined with the application of Carson's equations. Finally, the Kron reduction method is used to produce the final matrix that can be plugged into the unified branch model.

In IEEE distribution test feeders there are three spacing models for overhead lines listed in Table 2.1. Figure 2.5 shows the geometrical configuration of the conductors for the spacings listed. Another necessary piece of information is the conductor data. Table B.1 lists the characteristics of the various conductors that are used for the overhead configurations in distribution feeders.

The topology of the IEEE 13-bus feeder is included in Appendix A. Table A.1 lists the topology data. By matching this data with data in Table B.1, we extract the relevant GMR and resistance for each line segment. These two values are then used in Eq. 2.34 to determine the self impedance of the conductor section. Mutual impedances are determined by matching the geometrical configuration of the line segment with the configuration in Fig. 2.5, calculating the distance between the conductors and applying it to Eq. 2.36. For example, the calculation of the primitive series-impedance matrix for the line segment 632–633 is as follows:

Table 2.1 Overhead line spacings

Spacing ID	Type
500	Three-phase, 4 wire
505	Two-phase, 3 wire
510	Single-phase, 2 wire

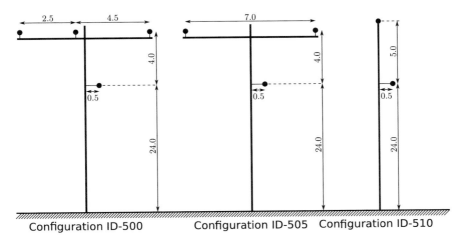

Fig. 2.5 Overhead line spacings

$$
\hat{\mathbf{z}}_{632-633} =
\begin{bmatrix}
\hat{z}_{aa} & \hat{z}_{ab} & \hat{z}_{ac} & \hat{z}_{an} \\
\hat{z}_{ba} & \hat{z}_{bb} & \hat{z}_{bc} & \hat{z}_{bn} \\
\hat{z}_{ca} & \hat{z}_{cb} & \hat{z}_{cc} & \hat{z}_{cn} \\
\hline
\hat{z}_{na} & \hat{z}_{nb} & \hat{z}_{nc} & \hat{z}_{nn}
\end{bmatrix}
\tag{2.38}
$$

$$
=
\left[
\begin{array}{ccc|c}
0.6873 + 1.5465j & 0.0953 + 0.7802j & 0.0953 + 0.8515j & 0.0953 + 0.7865j \\
0.0953 + 0.7802j & 0.6873 + 1.5465j & 0.0953 + 0.7266j & 0.0953 + 0.7674j \\
0.0953 + 0.8515j & 0.0953 + 0.7266j & 0.6873 + 1.5465j & 0.0953 + 0.7524j \\
\hline
0.0953 + 0.7865j & 0.0953 + 0.7674j & 0.0953 + 0.7524j & 0.6873 + 1.5465j
\end{array}
\right]
\tag{2.39}
$$

where the first three rows/columns represent phase conductors, and the fourth row/column represents the neutral conductor. The obtained matrix is reduced to a 3 by 3 matrix using the Kron reduction technique [19]:

$$
\mathbf{z}_{abc} = \hat{\mathbf{z}}_{ij} - \hat{\mathbf{z}}_{in} \cdot \hat{\mathbf{z}}_{nn}^{-1} \cdot \hat{\mathbf{z}}_{nj}
\tag{2.40}
$$

where

$$
\hat{\mathbf{z}}_{ij} =
\begin{bmatrix}
0.6873 + 1.5465j & 0.0953 + 0.7802j & 0.0953 + 0.8515j \\
0.0953 + 0.7802j & 0.6873 + 1.5465j & 0.0953 + 0.7266j \\
0.0953 + 0.8515j & 0.0953 + 0.7266j & 0.6873 + 1.5465j
\end{bmatrix},
\tag{2.41}
$$

$$\hat{z}_{in} = \begin{bmatrix} 0.0953 + 0.7865j \\ 0.0953 + 0.7674j \\ 0.0953 + 0.7524j \end{bmatrix}, \tag{2.42}$$

$$\hat{z}_{nj} = \begin{bmatrix} 0.0953 + 0.7865j & 0.0953 + 0.7674j & 0.0953 + 0.7524j \end{bmatrix}, \tag{2.43}$$

$$\hat{z}_{nn} = \begin{bmatrix} 0.6873 + 1.5465j \end{bmatrix}. \tag{2.44}$$

The series-impedance matrix of branch 632–633 is thus

$$\mathbf{z}_{632-633} = \begin{bmatrix} 0.7526 + 1.1814j & 0.1580 + 0.4237j & 0.1559 + 0.5017j \\ 0.1580 + 0.4237j & 0.7475 + 1.1983j & 0.1535 + 0.3850j \\ 0.1559 + 0.5017j & 0.1535 + 0.3850j & 0.7435 + 1.2113j \end{bmatrix}. \tag{2.45}$$

Line segments that have two-phase or single-phase configurations produce matrices where the missing conductors are zeroed out.

2.3.2.3 Shunt-Admittance Matrices of Overhead Lines in IEEE Distribution Test Feeders

The process of developing the shunt-admittance matrix for the overhead line is as follows. First, self and mutual potential coefficients are determined from the following equations:

$$\hat{P}_{ii} = 11.17689 \cdot \ln \frac{S_{ii}}{RD_i} \, \text{mile}/\mu F, \tag{2.46}$$

$$\hat{P}_{ij} = 11.17689 \cdot \ln \frac{S_{ij}}{D_{ij}} \, \text{mile}/\mu F, \tag{2.47}$$

where

- S_{ii} …is the distance from conductor i to its image in the ground i',
- S_{ij} …is the distance from conductor i to image in the ground of conductor j,
- D_{ij} …is the distance from conductor i to conductor j,
- RD_i …is the radius of conductor i.

note that the values of S_{ii}, RD_i, D_{ij}, and S_{ij} must all be in the same units. Next, the primitive potential coefficient matrix is formed. For a four-wire grounded-wye line, the primitive coefficient matrix will be of the form

$$\hat{\mathbf{P}}_{\text{primitive}} = \begin{bmatrix} \hat{P}_{aa} & \hat{P}_{ab} & \hat{P}_{ac} & \hat{P}_{an} \\ \hat{P}_{ba} & \hat{P}_{bb} & \hat{P}_{bc} & \hat{P}_{bn} \\ \hat{P}_{ca} & \hat{P}_{cb} & \hat{P}_{cc} & \hat{P}_{cn} \\ \hline \hat{P}_{na} & \hat{P}_{nb} & \hat{P}_{nc} & \hat{P}_{nn} \end{bmatrix} = \begin{bmatrix} \hat{\mathbf{P}}_{ij} & \hat{\mathbf{P}}_{in} \\ \hat{\mathbf{P}}_{nj} & \hat{\mathbf{P}}_{nn} \end{bmatrix} \tag{2.48}$$

In the final step, since the neutral conductor is grounded, the Kron reduction method is used to produce a matrix of size *Number of phases × Number of phases*:

$$\mathbf{P}_{abc} = \hat{\mathbf{P}}_{ij} - \hat{\mathbf{P}}_{in} \cdot \hat{\mathbf{P}}_{nn}^{-1} \cdot \hat{\mathbf{P}}_{nj} \tag{2.49}$$

The inverse of the potential coefficient matrix will give the *Number of phases × Number of phases* capacitance matrix:

$$\mathbf{C}_{abc} = \mathbf{P}_{abc}^{-1}. \tag{2.50}$$

For a two-phase line configuration, the capacitance matrix Eq. 2.50 will be 2 × 2 elements. Similarly, for a single-phase line configuration, the capacitance matrix will result in a single element. Normally, the shunt conductance is neglected and, thus, the shunt-admittance matrix is given by

$$\mathbf{Y}_{km,sh} = j \cdot 2\pi f \cdot \mathbf{C}_{abc} \tag{2.51}$$

To insert the obtained shunt-admittance matrix into the unified branch model, the matrix needs to be 3 × 3 elements. In the case of two-phase or shingle-phase line configurations, the row/column corresponding to the missing phase(s) needs to be zeroed out.

2.3.2.4   Series-Impedance Matrices of Underground Cables in IEEE Distribution Test Feeders

The process of developing the series-impedance matrix for underground cables is very similar to the process for overhead lines described in Sect. 2.3.2.2. The differences are in the geometric configurations of the cables and in the cables themselves.

In IEEE distribution test feeders, there are two cable spacings for the underground cables. They are listed in Table 2.2 and depicted in Fig. 2.6.

There are two popular types of underground cables, i.e., the concentric neutral cable, depicted in Fig. 2.7, and the tape-shielded cable, depicted in Fig. 2.8. In order to apply the modified Carson's equations, the resistance, and GMR of the phase conductor and the equivalent neutral must be determined. To compute the resistance and

Table 2.2 Underground cable spacings

Spacing ID	Type
515	Three-phase, 3 cable
520	Two-phase, 2 cable

Fig. 2.6 Underground cable spacings

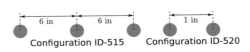

Configuration ID-515   Configuration ID-520

Fig. 2.7 Concentric neutral
cable

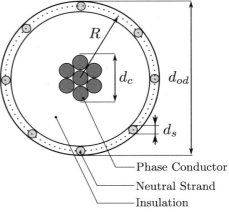

Phase Conductor

Neutral Strand

Insulation

Fig. 2.8 Tape-shielded
cable

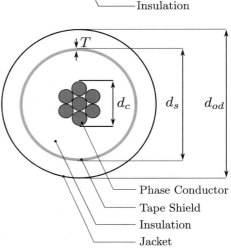

Phase Conductor

Tape Shield

Insulation

Jacket

GMR, the following data needs to be extracted from the datasheet of the underground
cable:

- d_c ...is the phase conductor diameter in inches,
- d_{od} ...is the nominal diameter over the concentric neutrals of the cable in inches,
- d_s ...is the diameter of a concentric neutral strand in inches,
- GMR_c ...is the geometric mean radius of the phase conductor in feet,
- GMR_s ...is the geometric mean radius of a neutral strand in feet,
- r_c ...is the resistance of the phase conductor in Ω/mile,
- r_s ...is the resistance of a solid neutral strand in Ω/mile,
- k ...is the number of concentric neutral strands.

The equivalent geometric mean radius of the concentric neutral cable is calculated
using

$$GMR_{cn} = \sqrt[k]{GMR_s \cdot k \cdot R^{k-1}} \tag{2.52}$$

where R is the radius of a circle passing through the center of the concentric neutral strands and is determined as

$$R = \frac{d_{od} - d_s}{24} \tag{2.53}$$

The equivalent resistance of the concentric neutral is determined using

$$r_{cn} = \frac{r_s}{k} \tag{2.54}$$

The spacings between the concentric neutral and phase conductors and other concentric neutrals in the various configurations are as follows:

- Concentric neutral to its own phase conductor (see Fig. 2.7) $D_{ij} = R$
- Concentric neutral to an adjacent concentric neutral $D_{ij} =$ center-to-center distance of the phase conductors.
- Concentric neutral to an adjacent phase conductor. The geometric mean distance between a concentric neutral and an adjacent phase conductor is given by

$$D_{ij} = \sqrt[k]{D_{nm}^k - R^k} \tag{2.55}$$

where D_{nm} is the center-to-center distance between the phase conductors. For configurations where the distance between the cables is much greater than the radius R, the expression can be simplified to $D_{ij} = D_{nm}$.

In applying the modified Carson's equations, the numbering of the phase and neutral conductors is important when constructing the primitive series-impedance matrix. For example, a three-phase, three-cable configuration depicted in Fig. 2.6 must be numbered as

- 1 is the phase conductor no. 1,
- 2 is the phase conductor no. 2,
- 3 is the phase conductor no. 3,
- 4 is the neutral of conductor no. 1,
- 5 is the neutral of conductor no. 2,
- 6 is the neutral of conductor no. 3.

the primitive series-impedance matrix is thus

$$\hat{\mathbf{z}}_{\text{primitive}} = \left[\begin{array}{ccc|ccc} \hat{z}_{11} & \hat{z}_{12} & \hat{z}_{13} & \hat{z}_{14} & \hat{z}_{15} & \hat{z}_{16} \\ \hat{z}_{21} & \hat{z}_{22} & \hat{z}_{23} & \hat{z}_{24} & \hat{z}_{25} & \hat{z}_{26} \\ \hat{z}_{31} & \hat{z}_{32} & \hat{z}_{33} & \hat{z}_{34} & \hat{z}_{35} & \hat{z}_{36} \\ \hline \hat{z}_{41} & \hat{z}_{42} & \hat{z}_{43} & \hat{z}_{44} & \hat{z}_{45} & \hat{z}_{46} \\ \hat{z}_{51} & \hat{z}_{52} & \hat{z}_{53} & \hat{z}_{54} & \hat{z}_{55} & \hat{z}_{56} \\ \hat{z}_{61} & \hat{z}_{62} & \hat{z}_{63} & \hat{z}_{64} & \hat{z}_{65} & \hat{z}_{66} \end{array}\right] = \left[\begin{array}{c|c} \hat{\mathbf{z}}_{ij} & \hat{\mathbf{z}}_{in} \\ \hline \hat{\mathbf{z}}_{nj} & \hat{\mathbf{z}}_{nn} \end{array}\right] \tag{2.56}$$

with the application of the Kron reduction method (Eq. 2.40), the *number of phases* × *number of phases* series-impedance matrix is obtained.

Table B.2 lists the characteristics of the various concentric neutral underground cables that are present in the IEEE distribution test feeders.

The tape-shielded cables consist of a central phase conductor covered by a thin layer of nonmetallic, semiconducting screen to which the insulating material is bonded. The insulation is covered by a semiconducting insulation screen. The bare copper shield is helically applied around the insulation screen. An insulating jacket surrounds the tape shield. The relevant geometric parameters are

- d_c ...is the diameter of the phase conductor in inches,
- d_s ...is the outside diameter of the tape shield in inches,
- d_{od} ...is the outside diameter over the jacket in inches,
- T ...is the thickness of the copper tape shield in mils.

To calculate the elements of the primitive series-impedance matrix, modified Carson's equations are applied. The resistance and GMR of the phase conductor are found in Table B.1. The resistance of the tape shield is given by

$$r_{\text{shield}} = 7.9385 \cdot 10^8 \frac{\rho}{d_s \cdot T} \, \Omega/\text{mile} \tag{2.57}$$

The resistivity (ρ) must be expressed in Ω-meters at 50 °C. The outside diameter of the tape shield (d_s) must be in inches and the thickness of the tape shield (T) in mils.

The GMR of the tape shield is the radius of a circle passing through the middle of the shield and is given by

$$GMR_{shield} = \frac{d_s - \frac{T}{1000}}{24}. \tag{2.58}$$

The spacings between the tape shields and the conductors in various configurations are as follows:

- Tape shield to its own phase conductor: $D_{ij} = GMR_{\text{shield}}$ ...is the radius to midpoint of the shield in feet.
- Tape shield to an adjacent tape shield D_{ij} = center-to-center distance of the phase conductors in feet.
- Tape shield to an adjacent phase or neutral conductor D_{ij} = center-to-center distance between phase conductors in feet.

again, the Kron method is applied to reduce the primitive series-impedance matrix into a *number of phases* × *number of phases* matrix.

2.3.2.5 Shunt-Admittance Matrices of Underground Cables in IEEE Distribution Test Feeders

Most underground distribution lines consist of one or more tape-shielded or concentric neutral cables. The process of developing the shunt-admittance matrix for any of the two configurations is the same as the one described in Sect. 2.3.2.3. To form a primitive admittance matrix, self and mutual potential coefficients need to be determined from Eqs. 2.46 and 2.47.

In the concentric neutral cable, all the neutral strands are at ground potential. Neutral stranding acts as a Faraday cage, and confines the electric field created by the phase conductor within the boundary of the concentric neutral strands. This enables us to use the relative permittivity of the insulation material as the relative permittivity of the medium ϵ_r. The capacitance from phase to ground for this type of cable is given by

$$C_{ph} = \frac{2\pi\epsilon}{\ln\frac{R}{RD_c} - \frac{1}{k}\ln\frac{k \cdot RD_s}{R}} \tag{2.59}$$

where (see Fig. 2.7 for reference)

- $\epsilon = \epsilon_0\epsilon_r$ ...is the permittivity of the medium,
- ϵ_r ...is the relative permittivity of the medium,
- ϵ_0 ...is the permittivity of free space,
- $RD_c = \frac{d_c}{2}$,
- $RD_s = \frac{d_s}{2}$.

From Eq. 2.59, the cable shunt admittance is obtained using

$$\mathbf{Y}_{shunt} = \begin{bmatrix} 1 & 0 & 0 \\ 0 & 1 & 0 \\ 0 & 0 & 1 \end{bmatrix} \cdot j \cdot 2\pi f \cdot C_{ph} \tag{2.60}$$

In the tape-shielded cable, as with the concentric neutral, the electric field is confined to the insulation. The capacitance of the tape-shielded cable can be derived from the expression developed for the concentric neutral cable (Eq. 2.59). The tape shield can be thought of as an infinite number of neutral strands:

$$C_{ag} = \lim_{k \to \infty} C_{ph} = \frac{2\pi\epsilon}{\ln\frac{R}{RD_c}}. \tag{2.61}$$

As with the concentric neutral cable, the cable shunt-admittance matrix is obtained as

$$\mathbf{Y}_{shunt} = \begin{bmatrix} 1 & 0 & 0 \\ 0 & 1 & 0 \\ 0 & 0 & 1 \end{bmatrix} \cdot j \cdot 2\pi f \cdot C_{ag} \tag{2.62}$$

2.3.3 Transformers

In modeling the transformers, we consider that the distribution three-phase transformers are mostly wound on a common core and, thus, all the windings are coupled to all the other windings. A two windings three-phase transformer then has a primitive or unconnected network consisting of six coupled coils. This type of transformer is depicted in Fig. 2.9.

From this basic representation, the primitive network can be derived, as shown in Fig. 2.10. The dotted lines represent the parasitic coupling between phases. This network is described by the primitive admittance matrix:

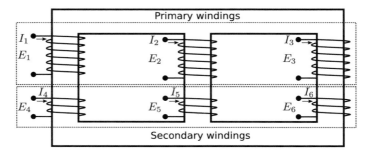

Fig. 2.9 Two-winding three-phase transformer

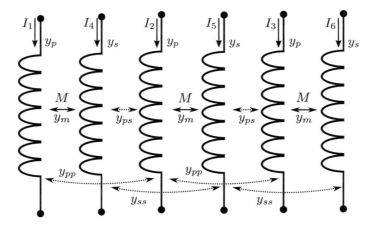

Fig. 2.10 Two-winding three-phase transformer

$$\begin{bmatrix} I_1 \\ I_2 \\ I_3 \\ I_4 \\ I_5 \\ I_6 \end{bmatrix} = \begin{bmatrix} y_{11} & y_{12} & y_{13} & y_{14} & y_{15} & y_{16} \\ y_{21} & y_{22} & y_{23} & y_{24} & y_{25} & y_{26} \\ y_{31} & y_{32} & y_{33} & y_{34} & y_{35} & y_{36} \\ y_{41} & y_{42} & y_{43} & y_{44} & y_{45} & y_{46} \\ y_{51} & y_{52} & y_{53} & y_{54} & y_{55} & y_{56} \\ y_{61} & y_{62} & y_{63} & y_{64} & y_{65} & y_{66} \end{bmatrix} \cdot \begin{bmatrix} E_1 \\ E_2 \\ E_3 \\ E_4 \\ E_5 \\ E_6 \end{bmatrix} \qquad (2.63)$$

Considering the reciprocal nature of the mutual couplings and assuming that the flux densities are symmetrically distributed between all the windings, Eq. 2.63 can be simplified to

$$\begin{bmatrix} I_1 \\ I_2 \\ I_3 \\ I_4 \\ I_5 \\ I_6 \end{bmatrix} = \begin{bmatrix} y_p & y_{pp} & y_{pp} & -y_m & y_{ps} & y_{ps} \\ y_{pp} & y_p & y_{pp} & y_{ps} & -y_m & y_{ps} \\ y_{pp} & y_{pp} & y_p & y_{ps} & y_{ps} & -y_m \\ -y_m & y_{ps} & y_{ps} & y_s & y_{ss} & y_{ss} \\ y_{ps} & -y_m & y_{ps} & y_{ss} & y_s & y_{ss} \\ y_{ps} & y_{ps} & -y_m & y_{ss} & y_{ss} & y_s \end{bmatrix} \cdot \begin{bmatrix} E_1 \\ E_2 \\ E_3 \\ E_4 \\ E_5 \\ E_6 \end{bmatrix} \qquad (2.64)$$

where

- y_p is the admittance of the primary coil,
- y_s is the admittance of the secondary coil,
- y_m is the mutual admittance between primary and secondary coils on the same core,
- y_{pp} is the mutual admittance between primary coils,
- y_{ps} is the mutual admittance between primary and secondary coils on different cores,
- y_{ss} is the mutual admittance between secondary coils.

For three separate single-phase units, the y_{pp}, y_{ps}, and y_{ss} are zero. In three-phase units, those values represent interphase coupling and can have a noticeable effect. However, for reasons of simplicity, these effects will be ignored in our analysis, and the three-phase units will be modeled as connected single-phase units. The simplifications produce the following model:

$$\begin{bmatrix} I_1 \\ I_2 \\ I_3 \\ I_4 \\ I_5 \\ I_6 \end{bmatrix} = \begin{bmatrix} y_p & 0 & 0 & M & 0 & 0 \\ 0 & y_p & 0 & 0 & M & 0 \\ 0 & 0 & y_p & 0 & 0 & M \\ M & 0 & 0 & y_s & 0 & 0 \\ 0 & M & 0 & 0 & y_s & 0 \\ 0 & 0 & M & 0 & 0 & y_s \end{bmatrix} \cdot \begin{bmatrix} E_1 \\ E_2 \\ E_3 \\ E_4 \\ E_5 \\ E_6 \end{bmatrix} = \mathbf{Y}_{\text{prim}} \cdot \mathbf{E}_{\text{prim}}. \qquad (2.65)$$

The network-admittance matrix for any two-winding three-phase transformer can be formed by the method of linear transformation. As an example, the formation of the admittance matrix for a grounded wye-delta connection is considered. The

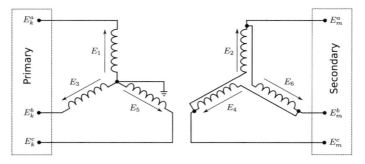

Fig. 2.11 The grounded wye-delta connection

transformer connection is illustrated in Fig. 2.11, and the connection matrix **C** relating the branch voltages to the node voltages is

$$
\begin{bmatrix} E_1 \\ E_2 \\ E_3 \\ E_4 \\ E_5 \\ E_6 \end{bmatrix} =
\begin{bmatrix}
1 & 0 & 0 & 0 & 0 & 0 \\
0 & 0 & 0 & 1 & -1 & 0 \\
0 & 1 & 0 & 0 & 0 & 0 \\
0 & 0 & 0 & 0 & 1 & -1 \\
0 & 0 & 1 & 0 & 0 & 0 \\
0 & 0 & 0 & -1 & 0 & 1
\end{bmatrix}
\begin{bmatrix} E_k^a \\ E_k^b \\ E_k^c \\ E_m^a \\ E_m^b \\ E_m^c \end{bmatrix} = \mathbf{C} \cdot \mathbf{E}_{\text{node}} \tag{2.66}
$$

The nodal admittance matrix $\mathbf{Y}_{\text{node}}$ is calculated by

$$
\mathbf{Y}_{\text{node}} = \mathbf{C}^T \cdot \mathbf{Y}_{\text{prim}} \cdot \mathbf{C}. \tag{2.67}
$$

If the primitive admittances are expressed in per unit, then the considered grounded wye-delta transformer must include an effective turns ratio $\sqrt{3}$. The upper-right and lower left quadrants of the matrix Y_{node} must be divided by $\sqrt{3}$ and the lower right quadrant by 3. Considering this, the nodal admittance model becomes

$$
\begin{bmatrix} I_k^a \\ I_k^b \\ I_k^c \\ I_m^a \\ I_m^b \\ I_m^c \end{bmatrix} =
\begin{bmatrix}
y & 0 & 0 & -y/\sqrt{3} & -y/\sqrt{3} & 0 \\
0 & y & 0 & 0 & -y/\sqrt{3} & y/\sqrt{3} \\
0 & 0 & y & y/\sqrt{3} & 0 & -y/\sqrt{3} \\
-y/\sqrt{3} & 0 & y/\sqrt{3} & \frac{2}{3}y & -\frac{1}{3}y & -\frac{1}{3}y \\
y/\sqrt{3} & -y/\sqrt{3} & 0 & -\frac{1}{3}y & \frac{2}{3}y & -\frac{1}{3}y \\
0 & y/\sqrt{3} & -y/\sqrt{3} & -\frac{1}{3}y & -\frac{1}{3}y & \frac{2}{3}y
\end{bmatrix}
\begin{bmatrix} E_k^a \\ E_k^b \\ E_k^c \\ E_m^a \\ E_m^b \\ E_m^c \end{bmatrix}. \tag{2.68}
$$

The models for the other common connections can be derived using the same method. In general, the transformer branch model can be represented by

$$
\begin{bmatrix} \mathbf{I}_k^{\{a,b,c\}} \\ \mathbf{I}_m^{\{a,b,c\}} \end{bmatrix} =
\begin{bmatrix} \mathbf{Y}_{pp} & \mathbf{Y}_{ps} \\ \mathbf{Y}_{sp} & \mathbf{Y}_{ss} \end{bmatrix} \cdot
\begin{bmatrix} \mathbf{E}_k^{\{a,b,c\}} \\ \mathbf{E}_m^{\{a,b,c\}} \end{bmatrix}, \tag{2.69}
$$

Table 2.3 Submatrices for common step-down transformer connections

Primary	Secondary	$\mathbf{Y}_{pp}$	$\mathbf{Y}_{ss}$	$\mathbf{Y}_{ps}$	$\mathbf{Y}_{sp}$
Y_g	Y_g	$\mathbf{Y}_I$	$\mathbf{Y}_I$	$-\mathbf{Y}_I$	$-\mathbf{Y}_I$
Y_g	Y	$\mathbf{Y}_{II}$	$\mathbf{Y}_{II}$	$-\mathbf{Y}_{II}$	$-\mathbf{Y}_{II}$
Y_g	Δ	$\mathbf{Y}_I$	$\mathbf{Y}_{II}$	$\mathbf{Y}_{III}$	$\mathbf{Y}_{III}^T$
Y	Y_g	$\mathbf{Y}_{II}$	$\mathbf{Y}_{II}$	$-\mathbf{Y}_{II}$	$-\mathbf{Y}_{II}$
Y	Y	$\mathbf{Y}_{II}$	$\mathbf{Y}_{II}$	$-\mathbf{Y}_{II}$	$-\mathbf{Y}_{II}$
Y	Δ	$\mathbf{Y}_{II}$	$\mathbf{Y}_{II}$	$\mathbf{Y}_{III}$	$\mathbf{Y}_{III}^T$
Δ	Y_g	$\mathbf{Y}_{II}$	$\mathbf{Y}_I$	$\mathbf{Y}_{III}$	$\mathbf{Y}_{III}^T$
Δ	Y	$\mathbf{Y}_{II}$	$\mathbf{Y}_{II}$	$\mathbf{Y}_{III}$	$\mathbf{Y}_{III}^T$
Δ	Δ	$\mathbf{Y}_{II}$	$\mathbf{Y}_{II}$	$-\mathbf{Y}_{II}$	$-\mathbf{Y}_{II}$

and it should be noted that the following relationship holds

$$\mathbf{Y}_{sp} = \mathbf{Y}_{ps}^T. \tag{2.70}$$

In our model the parameters of all three phases are assumed to be balanced. The most common three-phase connection can be modeled using three basic submatrices:

$$\mathbf{Y}_I = \begin{bmatrix} 1 & 0 & 0 \\ 0 & 1 & 0 \\ 0 & 0 & 1 \end{bmatrix} y, \tag{2.71}$$

$$\mathbf{Y}_{II} = \frac{1}{3} \begin{bmatrix} 2 & -1 & -1 \\ -1 & 2 & -1 \\ -1 & -1 & 2 \end{bmatrix} y, \tag{2.72}$$

$$\mathbf{Y}_{III} = \frac{1}{\sqrt{3}} \begin{bmatrix} -1 & 1 & 0 \\ 0 & -1 & 1 \\ 1 & 0 & -1 \end{bmatrix} y, \tag{2.73}$$

where y is the specified transformer-leakage admittance. Tables 2.3 and 2.4 list the combinations of $\mathbf{Y}_I$, $\mathbf{Y}_{II}$, and $\mathbf{Y}_{III}$ for the most common step-up and step-down transformer connections, respectively. It should be noted that off-nominal tap ratios are not considered in the node-admittance matrix. They must be modeled separately in the branch as $\mathbf{T}_{km}$ and $\mathbf{T}_{mk}$ matrices.

2.3.4 Voltage Regulators

As the loads on the distribution feeders vary, so do the feeder voltages. Voltage regulators are installed in distribution systems in order to control the voltage so that

Table 2.4 Submatrices for common step-up transformer connections

Primary	Secondary	$\mathbf{Y}_{pp}$	$\mathbf{Y}_{ss}$	$\mathbf{Y}_{ps}$	$\mathbf{Y}_{sp}$
Y_g	Y_g	$\mathbf{Y}_I$	$\mathbf{Y}_I$	$-\mathbf{Y}_I$	$-\mathbf{Y}_I$
Y_g	Y	$\mathbf{Y}_{II}$	$\mathbf{Y}_{II}$	$-\mathbf{Y}_{II}$	$-\mathbf{Y}_{II}$
Y_g	Δ	$\mathbf{Y}_I$	$\mathbf{Y}_{II}$	$\mathbf{Y}_{III}^T$	$\mathbf{Y}_{III}$
Y	Y_g	$\mathbf{Y}_{II}$	$\mathbf{Y}_{II}$	$-\mathbf{Y}_{II}$	$-\mathbf{Y}_{II}$
Y	Y	$\mathbf{Y}_{II}$	$\mathbf{Y}_{II}$	$-\mathbf{Y}_{II}$	$-\mathbf{Y}_{II}$
Y	Δ	$\mathbf{Y}_{II}$	$\mathbf{Y}_{II}$	$\mathbf{Y}_{III}^T$	$\mathbf{Y}_{III}$
Δ	Y_g	$\mathbf{Y}_{II}$	$\mathbf{Y}_I$	$\mathbf{Y}_{III}^T$	$\mathbf{Y}_{III}$
Δ	Y	$\mathbf{Y}_{II}$	$\mathbf{Y}_{II}$	$\mathbf{Y}_{III}^T$	$\mathbf{Y}_{III}$
Δ	Δ	$\mathbf{Y}_{II}$	$\mathbf{Y}_{II}$	$-\mathbf{Y}_{II}$	$-\mathbf{Y}_{II}$

every customer's voltage remains within an acceptable limit. The inner workings of the regulator and different methods of compensation are widely discussed in [19, 25]. In this subsection, only the relationship between the input and output quantities is of interest. The fundamental concepts of a single-phase voltage regulator will be reviewed in Sect. 2.3.4.1 in order to derive the branch-admittance matrices for the various three-phase connections in Sect. 2.3.4.2

2.3.4.1 Fundamental Concepts

Voltage regulator models can be derived from autotransformer models. An autotransformer is constructed with a two-winding transformer and connecting the two windings in series. The two possible connections, i.e., "buck" and "boost", are depicted in Figs. 2.12 and 2.13, respectively [25]. The two single-phase voltage regulators derived from these two possible connections are depicted in Figs. 2.14 and 2.15. The depicted situations represent the extreme "buck" and "boost" positions on a standard single-phase regulator. The resulting winding ratios for these configurations are $\frac{N_2 \pm N_1}{N_1}$. The basic equation relating the primary and secondary voltage sides is

$$\frac{E_P}{E_S} = \frac{E_A - E_b}{E_a - E_b} = a_{ab}. \tag{2.74}$$

From this equation, the $\mathbf{T}_{km}$ and $\mathbf{T}_{mk}$ matrices of the unified branch model can be derived for any possible three-phase connection of the voltage regulator.

2.3.4.2 Three-Phase Connections

A bank of single-phase regulators connected in the grounded-wye configuration is shown in Figs. 2.16. Each of the three regulators controls the voltage delivered to

Fig. 2.12 Bucking autotransformer configuration

Fig. 2.13 Boosting autotransformer configuration

Fig. 2.14 Bucking single-phase voltage regulator

Fig. 2.15 Boosting single-phase voltage regulator

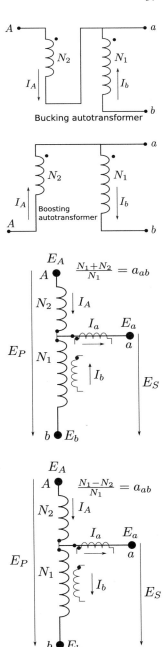

Fig. 2.16 Single-phase
regulators connected in
grounded-wye

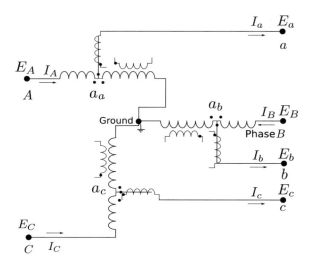

the customers on each phase. The regulators act independently, based on the loading
and impedance of an individual feeder phase. Each regulator can regulate the voltage
across its full range. The standard single-phase voltage regulator has 32 steps of
regulation and a $\pm 10\%$ range. The steps are symmetrically distributed around zero;
therefore, for a total range of $\pm 10\%$, 32 steps yield a $\frac{5}{8}\%$ step size.

If we consider Eq. 2.74, the relationship between the primary and secondary side
for the wye-connected single-phase regulator is

$$
\begin{bmatrix} E_A \\ E_B \\ E_C \end{bmatrix} = \begin{bmatrix} a_a & 0 & 0 \\ 0 & a_b & 0 \\ 0 & 0 & a_c \end{bmatrix} \cdot \begin{bmatrix} E_a \\ E_b \\ E_c \end{bmatrix} = \mathbf{T} \cdot \mathbf{E}^{\{a,b,c\}}. \tag{2.75}
$$

The line-to-neutral voltages are just scaled with the appropriate regulator ratio.

A bank of delta-connected regulators in boost configuration is depicted in
Fig. 2.17. The voltage that is applied to the control is proportional to the line-to-
line voltage across the load. Again, considering Eq. 2.74, the relationship between
line-to-neutral primary and secondary voltages is

$$
\begin{bmatrix} E_A \\ E_B \\ E_C \end{bmatrix} = \begin{bmatrix} a_{ab} & (1 - a_{ab}) & 0 \\ 0 & a_{bc} & (1 - a_{bc}) \\ (1 - a_{ca}) & 0 & a_{ca} \end{bmatrix} \cdot \begin{bmatrix} E_a \\ E_b \\ E_c \end{bmatrix} = \mathbf{T} \cdot \mathbf{E}^{\{a,b,c\}}. \tag{2.76}
$$

Figure 2.18 depicts the primary and secondary line-to-line and line-to-neutral phasors
for the closed delta-connected voltage regulator with the following tap ratios: $a_{ca} =
0.93$, $a_{bc} = 0.9$, and $a_{ab} = 1.1$. It is obvious that besides magnitude scaling, phase
shifts also occur. It should also be noted that the range of regulation in terms of the
line-to-neutral voltages is less than $\pm 10\%$ (the range of one single-phase voltage
regulator unit).

Fig. 2.17 Single-phase regulators in a closed delta connection

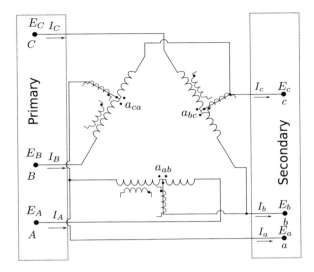

Fig. 2.18 Phase diagram for closed delta connection

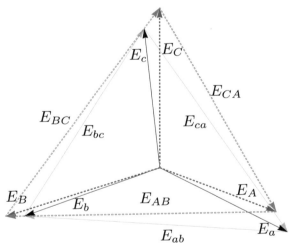

A bank of open-delta-connected single-phase voltage regulators is depicted in Fig. 2.19. Considering Eq. 2.74, the relationships between the primary and secondary voltages are derived

$$
\begin{bmatrix} E_A \\ E_B \\ E_C \end{bmatrix} = \begin{bmatrix} a_{ab} & (1-a_{ab}) & 0 \\ 0 & 1 & 0 \\ 0 & (1-a_{cb}) & a_{cb} \end{bmatrix} \cdot \begin{bmatrix} E_a \\ E_b \\ E_c \end{bmatrix} = \mathbf{T} \cdot \mathbf{E}^{\{a,b,c\}}. \tag{2.77}
$$

Figure 2.20 depicts the primary and secondary line-to-line and line-to-neutral phasors for the open-delta-connected voltage regulator.

Fig. 2.19 Two single-phase regulators in an open-delta connection

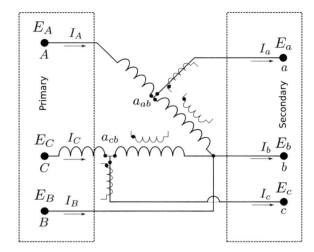

Fig. 2.20 Phase diagram for an open-delta connection

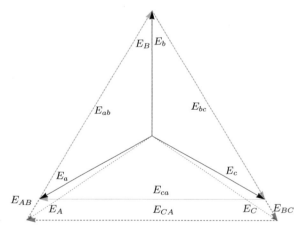

2.3.5 The Complete Network Model

To construct a complete network model, a system-admittance matrix is needed. With branch matrices for possible components developed, the construction of a system-admittance matrix is a straightforward process. The proposed unified branch model is applied to a fictitious 4-bus system shown in Fig. 2.21. The example includes a transformer in a delta-grounded wye connection, a conductor section with a voltage regulator in an open-delta connection, and an additional conductor section. The following example covers the use of the unified branch model for the majority of practical scenarios. Other possible cases can be inferred from the demonstrated ones. The three example branches have the following branch matrices:

Fig. 2.21 Example system

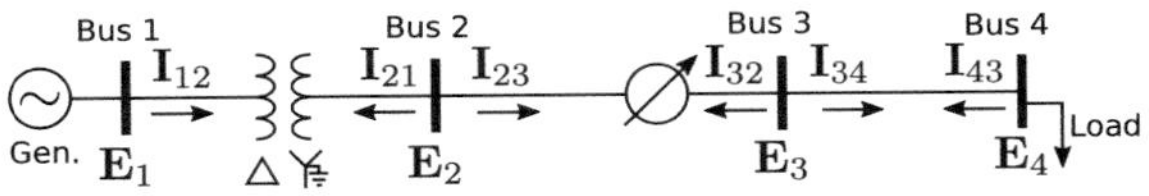

- *The 1–2 branch*: The transformer-leakage admittance matrix is constructed from the basic submatrices in Table 2.4. For the transformer in the example, the submatrices are $\mathbf{Y}_{pp} = \mathbf{Y}_{II}$, $\mathbf{Y}_{pq} = \mathbf{Y}_{III}$, $\mathbf{Y}_{qp} = \mathbf{Y}_{III}^T$, and $\mathbf{Y}_{qq} = \mathbf{Y}_{I}$. Possible off-nominal tap ratios are considered in the $\mathbf{T}_{12}$ and $\mathbf{T}_{21}$ matrices. The complete branch-admittance matrix is

$$\mathbf{Y}_{b,12} = \begin{bmatrix} \mathbf{T}_{12} \cdot (\mathbf{Y}_{pp} + \mathbf{Y}_{km,sh}) \cdot \mathbf{T}_{12} & -\mathbf{T}_{12} \cdot \mathbf{Y}_{pq} \cdot \mathbf{T}_{21} \\ -\mathbf{T}_{21} \cdot \mathbf{Y}_{qp} \cdot \mathbf{T}_{12} & \mathbf{T}_{21} \cdot (\mathbf{Y}_{qq} + \mathbf{Y}_{mk,sh}) \cdot \mathbf{T}_{21} \end{bmatrix} \quad (2.78)$$

- *The 2–3 branch*: The branch consists of a conductor section and a voltage regulator at the side of bus 3. The resulting branch matrix is

$$\mathbf{Y}_{b,23} = \begin{bmatrix} (\mathbf{Y}_{23} + \mathbf{Y}_{23,sh}) & \mathbf{Y}_{23} \cdot \mathbf{T}_{32} \\ -\mathbf{T}_{32} \cdot \mathbf{Y}_{23} & \mathbf{T}_{32} \cdot (\mathbf{Y}_{23} + \mathbf{Y}_{23,sh}) \cdot \mathbf{T}_{32} \end{bmatrix} \quad (2.79)$$

- *The 3–4 branch*: The branch consists of a conductor section only. The resulting branch matrix is

$$\mathbf{Y}_{b,34} = \begin{bmatrix} (\mathbf{Y}_{34} + \mathbf{Y}_{34,sh}) & -\mathbf{Y}_{34} \\ -\mathbf{Y}_{34} & (\mathbf{Y}_{34} + \mathbf{Y}_{43,sh}) \end{bmatrix} \quad (2.80)$$

The system-admittance matrix is constructed from the developed branch matrices. Branch submatrices are added to the corresponding elements in the system matrix. The final result (superscripts indicate the inner submatrix of every branch matrix) is

$$\mathbf{Y}_{sys} = \begin{bmatrix} \mathbf{Y}_{b,12}^{11} & \mathbf{Y}_{b,12}^{12} & & \\ \mathbf{Y}_{b,12}^{21} & \mathbf{Y}_{b,12}^{22} + \mathbf{Y}_{b,23}^{11} & \mathbf{Y}_{b,23}^{12} & \\ & \mathbf{Y}_{b,23}^{21} & \mathbf{Y}_{b,23}^{22} + \mathbf{Y}_{b,34}^{11} & \mathbf{Y}_{b,34}^{12} \\ & & \mathbf{Y}_{b,34}^{21} & \mathbf{Y}_{b,34}^{22} \end{bmatrix}. \quad (2.81)$$

With the system-admittance matrix available, Eqs. 2.82 and 2.83 show expressions for the bus power injections. G_{km}^{ap} and B_{km}^{ap}, respectively, represent the real and imaginary components of the (a, p) and (k, m) system-admittance submatrix, and Λ is the set of buses adjacent to bus k, including bus k.

$$P_k^a = |V_k^a| \sum_{m \in \Lambda} \sum_{p \in \Omega_p} |V_m^p| (G_{km}^{ap} \cos(\phi_k^a - \phi_m^p) + B_{km}^{ap} \sin(\phi_k^a - \phi_m^p)) \quad (2.82)$$

$$Q_k^a = |V_k^a| \sum_{m \in \Lambda} \sum_{p \in \Omega_p} |V_m^p| (G_{km}^{ap} \sin(\phi_k^a - \phi_m^p) - B_{km}^{ap} \cos(\phi_k^a - \phi_m^p)) \quad (2.83)$$

References

1. F.C. Schweppe, Power system static-state estimation part I: the exact model. IEEE Trans. Power Appar. Syst. (1970)
2. F.C. Schweppe, B. Rom Douglas, Power system static-state estimation part II: the approx. model. IEEE Trans. Power Appar. Syst. (1970)
3. F.C. Schweppe, Power system static-state estimation, part III: implementation. IEEE Trans. Power Appar. Syst. **1**, 130–135 (1970). Accessed 06 Oct 2014
4. U. Kuhar, Three-phase state estimation in power distribution systems. Ph.D. Thesis. Jožef Stefan International Postgraduate School Ljubljana, Slovenia, 2018
5. D. Della Giustina et al., Electrical distribution system state estimation: measurement issues and challenges. IEEE Instrum. Meas. Mag. (2014). 1094.6969/14
6. C.N. Lu, J.H. Teng, W.-H. Liu, Distribution system state estimation. IEEE Trans. Power Syst. **10**(1), 229–240 (1995). Accessed 07 Dec 2014
7. M.E. Baran, A.W. Kelley, A branch-current-based state estimation method for distribution systems. IEEE Trans. Power Syst. **10**(1), 483–491 (1995)
8. K. Li, State estimation for power distribution system and measurement impacts. IEEE Trans. Power Syst. **11**(2), 911–916 (1996). Accessed 24 Aug 2015
9. A.P. Sakis Meliopoulos, F. Zhang, Multiphase power flow and state estimation for power distribution systems. IEEE Trans. Power Syst. **11**(2), 939–946 (1996)
10. M.E. Baran, A.W. Kelley, State estimation for real-time monitoring of distribution systems. IEEE Trans. Power Syst. **9**(3), 1601–1609 (1994). Accessed 07 Dec 2014
11. U. Kuhar et al., A unified three-phase branch model for a distribution system state estimation, in *2016 IEEE PES Innovative Smart Grid Technologies Conference Europe (ISGT-Europe)* (IEEE, 2016), pp. 1–6
12. J.J. Grainger, W.D. Stevenson, *Power System Analysis*. McGraw-Hill Series in Electrical and Computer Engineering: Power and Energy (McGraw-Hill, New York, 1994). ISBN: 9780070612938
13. A. Monticelli, A. Garcia, Fast decoupled state estimator. In: ()
14. R. Singh, B.C. Pal, R.A. Jabr, Distribution system state estimation through Gaussian mixture model of the load as pseudo-measurement. en. IET Gener. Transm. Distrib. **4**(1), 50 (2010). ISSN: 17518687. https://doi.org/10.1049/iet-gtd.2009.0167. Accessed 06 Dec 2014
15. E. Manitsas et al., Distribution system state estimation using an artificial neural network approach for pseudo measurement modeling. IEEE Trans. Power Syst. **27**(4), 1888–1896 (2012). ISSN: 0885-8950, 1558-0679. https://doi.org/10.1109/TPWRS.2012.2187804. Accessed 20 May 2015
16. S.M. Kay, *Fundamentals of Statistical Signal Processing: Estimation Theory* (Prentice-Hall, Inc., Upper Saddle River, 1993). ISBN: 0-13- 345711-7
17. A. Abur, A.G. Exposito, *Power System State Estimation: Theory and Implementation* (Marcel Dekker, New York, 2004). ISBN: 978-0-8247-5570-6
18. L. Mili et al., Robust state estimation based on projection statistics. IEEE Trans. Power Syst. **11**(2), 1118–1127 (1996)
19. W.H. Kersting, *Distribution System Modeling and Analysis*. The Electric Power Engineering Series (CRC Press, Boca Raton, 2002). ISBN: 0-8493-0812-7
20. P. Xiao, D.C. Yu, W. Yan, A unified three-phase transformer model for distribution load flow calculations. en. IEEE Trans. Power Syst. **21**(1), 153–159 (2006). ISSN: 0885-8950. https://doi.org/10.1109/TPWRS.2005.857847. Accessed 25 May 2015
21. R. Hoffman, Practical state estimation for electric distribution networks, in *2006 IEEE PES Power Systems Conference and Exposition, PSCE'06* (IEEE, 2006), pp. 510–517
22. P. Janssen, T. Sezi, J.-C. Maun, Distribution system state estimation using unsynchronized phasor measurements, in *2012 3rd IEEE PES International Conference and Exhibition on Innovative Smart Grid Technologies (ISGT Europe)* (IEEE, 2012), pp. 1–6
23. D.A. Haughton, G.T. Heydt, A linear state estimation formulation for smart distribution systems. IEEE Trans. Power Syst. **28**(2), 1187–1195 (2013)

24. A. Monticelli, *State Estimation in Electric Power Systems: A Generalized Approach*. The Kluwer International Series in Engineering and Computer Science; Power Electronics and Power Systems SECS 507 (Kluwer Academic Publishers, Boston, 1999). ISBN: 978-0-7923-8519-6
25. M.T. Bishop, J.D. Foster, D.A. Down, The application of single-phase voltage regulators on three-phase distribution systems, in *Rural Electric Power Conference, 1994. Papers Presented at the 38th Annual Conference* (IEEE, 1994), pp. C2–1

Chapter 3
Numerical Solution of the Estimation Problem

3.1 Introduction

The reviewed estimators, MLE, WLS, LAV, and SHGM, are solved by minimizing the objective function they yield (Eqs. 2.7, 2.10, 2.11, and 2.12). Since the objective function is generally nonlinear, there is no closed-form solution. In such a case, we need to resort to iterative methods. Typical examples include the Newton–Raphson method and the Gauss–Newton method. In general, these methods will converge on a global minimum if the initial guess is close to the true minimum. If not, convergence might not be attained, or the method could converge to a local minimum. The difficulty with the use of iterative methods is that, in general, we do not know beforehand if they will converge and whether the solution produced is the global minimum. A good initial guess is, thus, of paramount importance. In the following sections, the Newton–Raphson, Gauss–Newton, augmented Hachtel, and Iteratively Reweighted Least Squares numerical methods will be reviewed.

3.2 Newton–Raphson Iterative Method

The Newton–Raphson iterative numerical method can be used to obtain the minimum of the objective function that is produced by the WLS (and MLE) estimator. The algorithm can be derived by differentiating the objective function Eq. 2.10:

$$\frac{\partial J(\mathbf{x})_{WLS}}{\partial x_j} = -2 \sum_{i=1}^{m} r_i^{-1} \big(f_i - h_i(\mathbf{x}) \big) \frac{\partial h_i(\mathbf{x})}{\partial x_j} = 0, \qquad (3.1)$$

© The Author(s), under exclusive license to Springer Nature Switzerland AG 2020
U. Kuhar et al., *Observability of Power-Distribution Systems*,
SpringerBriefs in Applied Sciences and Technology,
https://doi.org/10.1007/978-3-030-39476-9_3

where $j = 1, 2, \ldots, n$. The Jacobian matrix with partial derivatives is defined

$$\left[\frac{\partial \mathbf{h}(\mathbf{x})}{\partial \mathbf{x}} \right]_{i,j} = \frac{\partial h_i(\mathbf{x})}{\partial x_j}, \tag{3.2}$$

where $i = 1, 2, \ldots, m$, $j = 1, 2, \ldots, n$. The necessary conditions in vector form are then

$$\frac{\partial J(\mathbf{x})_{WLS}}{\partial \mathbf{x}} = \frac{\partial \mathbf{h}(\mathbf{x})^T}{\partial \mathbf{x}} \mathbf{R}^{-1}(\mathbf{f} - \mathbf{h}(\mathbf{x})) = \mathbf{0}. \tag{3.3}$$

Equation 3.3 represents a set of m nonlinear equations. To obtain the solution, the iteration scheme is constructed as follows, and we define

$$\mathbf{g}(\mathbf{x}) = \frac{\partial \mathbf{h}(\mathbf{x})^T}{\partial \mathbf{x}} \mathbf{R}^{-1}(\mathbf{f} - \mathbf{h}(\mathbf{x})) \tag{3.4}$$

the (Newton–Raphson) iteration is then

$$\mathbf{x}_{k+1} = \mathbf{x}_k - \left(\frac{\partial \mathbf{g}(\mathbf{x}_k)}{\partial \mathbf{x}_k} \right)^{-1} \mathbf{g}(\mathbf{x}_k). \tag{3.5}$$

The Jacobian of $\mathbf{g}$ is

$$\frac{\partial \left[\mathbf{g}(\mathbf{x}) \right]_i}{\partial x_j} = \frac{\partial}{\partial x_j} \left[\sum_{p=1}^{m} r_p^{-1}(f_p - h_p(\mathbf{x})) \frac{\partial h_p(\mathbf{x})}{\partial x_i} \right] \tag{3.6}$$

$$= \sum_{p=1}^{m} \left[r_p^{-1}(f_p - h_p(\mathbf{x})) \frac{\partial^2 h_p(\mathbf{x})}{\partial x_i \partial x_j} - r_p^{-1} \frac{\partial h_p(\mathbf{x})}{\partial x_i} \frac{\partial h_p(\mathbf{x})}{\partial x_i} \right]. \tag{3.7}$$

In vector form, the Jacobian is

$$\frac{\partial \mathbf{g}(\mathbf{x})}{\partial \mathbf{x}} = \sum_{p=1}^{m} \mathbf{G}_p(\mathbf{x}) r_p^{-1}(f_p - h_p(\mathbf{x})) - \mathbf{H}^T(\mathbf{x}) \mathbf{R}^{-1} \mathbf{H}(\mathbf{x}) \tag{3.8}$$

where the matrices $\mathbf{H}(\mathbf{x})$, $\mathbf{G}_p(\mathbf{x})$ are

$$\left[\mathbf{H}(\mathbf{x}) \right]_{i,j} = \left[\frac{\partial \mathbf{g}(\mathbf{x})}{\partial \mathbf{x}} \right]_{i,j} = \frac{\partial h_i(\mathbf{x})}{\partial x_j}, \tag{3.9}$$

$$\left[\mathbf{G}_p(\mathbf{x}) \right]_{i,j} = \frac{\partial^2 h_i(\mathbf{x})}{\partial x_i \partial x_j}. \tag{3.10}$$

Considering Eqs. 3.5, 3.8, and 3.4, the final form of the Newton–Raphson iteration is

$$\mathbf{x}_{k+1} = \mathbf{x}_k + \left(\mathbf{H}^T(\mathbf{x}_k)\mathbf{R}^{-1}\mathbf{H}(\mathbf{x}_k) - \sum_{p=1}^{m} \mathbf{G}_p(\mathbf{x}_k)r_i^{-1}(f_p - h_p(\mathbf{x}_k)) \right)^{-1} \mathbf{H}^T(\mathbf{x}_k)\mathbf{R}^{-1}(\mathbf{f} - \mathbf{h}(\mathbf{x}_k)).$$

(3.11)

The main difficulty with the Newton–Raphson method is calculating the matrix of second derivatives $\mathbf{G}_p$ of the measurement functions. In general, the method has a quadratic rate of convergence. The convergence, however, is not guaranteed and depends heavily on the selection of the initial guess. In our implementation, for medium-sized distribution networks (IEEE distribution feeders up to 123 buses) with moderate loading, the selection of the flat-start (i.e., phase angles of an unloaded feeder and nominal voltage magnitudes), the initial solution was usually suitable. This initial solution needs to account for phase shifts due to transformer connections, and magnitude scaling due to tap changers (see Sect. 3.6). The measurement configurations with a prevailing number of power-flow measurements can have convergence problems with the flat-start initial solution; because real and reactive power flows are nearly zero at the flat-start, the Jacobian matrix can become ill-conditioned (or even singular) during the first iteration. One of the possible solutions for this condition is explained in [1].

Except for the possible convergence problems that can be solved with a better choice of the initial solution in the majority of cases, and the difficulty of computating the second derivatives of the measurement functions, the Newton–Raphson iteration efficiently delivers a solution for a state estimation problem.

3.3 Gauss–Newton Iteration Method

The Gauss–Newton method is a modification of the Newton–Raphson method. In comparison with the Newton–Raphson method, it does not require second-order derivatives of the measurement functions. The trade-off is that it has slower rate of convergence. The method can suffer from the same convergence problems if the initial solution is poorly selected or in the case where the measurement configuration contains a prevailing number of power-flow measurements.

The method can be derived as follows. The signal model is linearized about some nominal $\mathbf{x}$

$$\mathbf{h}_p(\mathbf{x}) = \mathbf{h}(\mathbf{x}_0) + \left. \frac{\partial \mathbf{h}_p(\mathbf{x})}{\partial \mathbf{x}} \right|_{\mathbf{x}=\mathbf{x}_0} (\mathbf{x} - \mathbf{x}_0)$$

(3.12)

the WLS error becomes

$$J = \sum_{p=1}^{m} r_i^{-1}(f_p - h_p(\mathbf{x}))^2$$

(3.13)

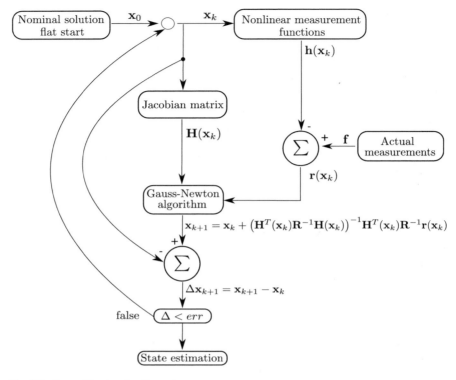

Fig. 3.1 Gauss–Newton iteration scheme

$$\approx \sum_{p=1}^{n} r_i^{-1} \left(f_p - h_p(\mathbf{x}_0) + \left.\frac{\partial h_p(\mathbf{x})}{\partial \mathbf{x}}\right|_{\mathbf{x}=\mathbf{x}_0} \mathbf{x}_0 - \left.\frac{\partial h_p(\mathbf{x})}{\partial \mathbf{x}}\right|_{\mathbf{x}=\mathbf{x}_0} \mathbf{x} \right)^2 \tag{3.14}$$

$$= \left(\mathbf{f} - \mathbf{h}(\mathbf{x}_0) + \mathbf{H}(\mathbf{x}_0)\mathbf{x}_0 - \mathbf{H}(\mathbf{x}_0)\mathbf{x}\right)^T \mathbf{R}^{-1} \left(\mathbf{f} - \mathbf{h}(\mathbf{x}_0) + \mathbf{H}(\mathbf{x}_0)\mathbf{x}_0 - \mathbf{H}(\mathbf{x}_0)\mathbf{x}\right). \tag{3.15}$$

The obtained linearized criterion is solved for $\mathbf{x}$ and the Gauss–Newton iteration scheme is obtained

$$\mathbf{x}_{k+1} = \mathbf{x}_k + \left(\mathbf{H}^T(\mathbf{x}_k)\mathbf{R}^{-1}\mathbf{H}(\mathbf{x}_k)\right)^{-1}\mathbf{H}^T(\mathbf{x}_k)\mathbf{R}^{-1}(\mathbf{f} - \mathbf{h}(\mathbf{x}_k)). \tag{3.16}$$

The Gauss–Newton iteration scheme is depicted in Fig. 3.1. It is evident that the Gauss–Newton iteration scheme is based on successive linearizations followed by solving a linear least squares problems. Linear sub-problems are solved with the use of the product $\mathbf{H}^T(\mathbf{x}_k)\mathbf{H}(\mathbf{x}_k)$, which maintains the sparsity of the original matrix, but is intrinsically ill-conditioned [2]. To this end, methods were developed that circumvent this difficulty, such as the augmented-matrix method, which is reviewed in Sect. 3.4.

3.4 Hachtel's Augmented-Matrix Method

The presence of the product of matrix $\mathbf{H}(\mathbf{x}_k)$ with its transposed version $\mathbf{H}^T(\mathbf{x}_k)$ in Eq. 3.16 has a significant effect on the conditioning of the system. The condition number is a measure of how much the output value of the system (or equation) can change for a small change in the input value. For a matrix, it is calculated as

$$\kappa(\mathbf{A}) = ||\mathbf{A}|| \times ||\mathbf{A}^{-1}|| \tag{3.17}$$

and it can be shown that

$$\kappa(\mathbf{A}^T \mathbf{A}) = \left[\kappa(\mathbf{A})^2\right]. \tag{3.18}$$

For condition numbers above 10^3, the system can be treated as ill-conditioned, meaning that to successfully compute the result through the normal equations the condition number should be smaller than that. Specific sources of ill-conditioning were identified in the literature [1]:

- Large weighting factors used along with virtual measurements,
- Short and long lines simultaneously present at the same bus,
- A large proportion of injection measurements.

Conditioning in the presence of the listed sources is further exacerbated as the system grows in size. The problem of solving the least squares problem must, as a result, be restated.

Several methods were developed to avoid or neutralize the listed sources of ill-conditioning. Methods using orthogonal transformations, such as those of Householder and Givens, have better stability properties [3, 4]. On the other hand, orthogonal transformations can give a large "fill-in" in sparse matrices [5]. A method that avoids this has been described in [6], but the operation count might be higher than with the normal equations. Based on a comparison of approaches with the normal equations, orthogonal transformations, Hachtel's method, and method of Peters and Wilkinson, Ref. [5] recommends the use of Hachtel's method for sparse WLS problems.

As the second and third identified sources are broader system-design factors that cannot be influenced by the selection of the estimator-solution method, the first source, i.e., the use of large weighting factors together with virtual measurements, can be eliminated with the implementation of Hachtel's method. Hachtel's method restates the WLS problem as an optimization problem where virtual measurements are presented as explicit optimization constraints. The problem is thus written as

$$\underset{\mathbf{x}}{\text{minimize}} \quad J_{WLS}(\mathbf{x}) = \frac{1}{2}(\mathbf{f} - \mathbf{h}(\mathbf{x}))^T \mathbf{R}^{-1}(\mathbf{f} - \mathbf{h}(\mathbf{x}))$$
$$\text{subject to} \quad \mathbf{c}(\mathbf{x}) = \mathbf{0} : \lambda, \tag{3.19}$$

where $J_{WLS}(\mathbf{x})$ was multiplied by $1/2$ because it simplifies the derivation. The multiplication does not change the objective function in any relevant way. The J_{WLS} is minimized iteratively as a sequence of linear problems, where in the kth iteration we minimize

$$J_{WLS}(\mathbf{x}) = \frac{1}{2}(\Delta\mathbf{f} - \mathbf{H}\Delta\mathbf{x})^T\mathbf{R}^{-1}(\Delta\mathbf{f} - \mathbf{H}\Delta\mathbf{x}),\tag{3.20}$$

where

$$\Delta\mathbf{x} = \mathbf{x}_k - \mathbf{x}_{k-1},\tag{3.21}$$

$$\Delta\mathbf{z} = \mathbf{z} - \mathbf{h}(\mathbf{x}_k),\tag{3.22}$$

$$\mathbf{x}_k = \text{solution in the kth iteration,}$$

$$\mathbf{H} = \text{Jacobian matrix evaluated at } \mathbf{x}_k.$$

In order to include the nonlinear equality constraints from Eq. 3.19 in a solution algorithm, they need to be linearized

$$\mathbf{C}\Delta\mathbf{x} = \Delta\mathbf{c}\tag{3.23}$$

where $\mathbf{C}$ is a o by n matrix of first derivatives:

$$[\mathbf{C}]_{i,j} = \left[\frac{\partial\mathbf{c}(\mathbf{x})}{\partial\mathbf{x}}\right]_{i,j} = \frac{\partial c_i(\mathbf{x})}{\partial x_j},\tag{3.24}$$

and $\Delta\mathbf{c} = -\mathbf{c}(\mathbf{x}_k)$ compensates for the linearization error in the previous iteration. The Lagrangian is then constructed as

$$\mathscr{L} = \frac{1}{2}(\Delta\mathbf{f} - \mathbf{H}\Delta\mathbf{x})^T\mathbf{R}^{-1}(\Delta\mathbf{f} - \mathbf{H}\Delta\mathbf{x}) - \boldsymbol{\lambda}^T(\Delta\mathbf{c} - \mathbf{C}\Delta\mathbf{x}).\tag{3.25}$$

The minimum is found by deriving with respect to $\mathbf{x}$:

$$\frac{\partial\mathscr{L}}{\partial\mathbf{x}} = \mathbf{H}^T(\Delta\mathbf{z} - \mathbf{H}\Delta\mathbf{x}) - \mathbf{C}^T\boldsymbol{\lambda} = \mathbf{0}.\tag{3.26}$$

The residual vector $\mathbf{r}$ is treated as an unknown

$$\mathbf{r} = \Delta\mathbf{z} - \mathbf{H}\Delta\mathbf{x}\tag{3.27}$$

Equations 3.23, 3.26, and 3.27 are assembled into a matrix equation:

$$\begin{bmatrix} \alpha\mathbf{R} & \mathbf{H} & \mathbf{0} \\ \mathbf{H}^T & \mathbf{0} & \mathbf{C}^T \\ \mathbf{0} & \mathbf{C} & \mathbf{0}^T \end{bmatrix} \cdot \begin{bmatrix} \alpha^{-1}\mathbf{r} \\ \Delta\mathbf{x} \\ -\alpha^{-1}\boldsymbol{\lambda} \end{bmatrix} = \begin{bmatrix} \Delta\mathbf{z} \\ \mathbf{0} \\ \Delta\mathbf{c} \end{bmatrix}\tag{3.28}$$

Equation 3.28 is an augmented-matrix formulation of the linear WLS problem with equality constraints. The system has $m + n + o$ unknowns, and so the problem dimension and thus the computational load are increased, compared to the Newton–Raphson and Gauss–Newton methods with n unknowns. However, the system remains sparse, and this property can be beneficially leveraged in the implementation. Conditioning of the system is improved significantly, as the computation of product $\mathbf{H}^T\mathbf{H}$ is avoided.

The factor α in Eq. 3.28 cancels out mathematically, but it influences the conditioning of the equation. In our case, we used the following value [1]:

$$\alpha = max \ R_{ii}^{-1}. \tag{3.29}$$

Comparison studies [1] show that, for a WLS estimator, Hachtel's method is the cheapest and best-conditioned approach for realistic power networks.

3.5 Iteratively Reweighted Least Squares

The solutions to the SHGM and LAV estimators can be obtained using different methods. The LAV estimator is actually a linear programming problem and can be solved using a plethora of methods discussed in the literature [7]. The Newton–Raphson method requires a computation of second derivatives of the $|f_- h_i(\mathbf{x})|$ and $\gamma_i^2 \rho(r_i)$ functions for the LAV and SHGM estimators, respectively. A method that does not require second-order derivatives and is suitable for the solution of both estimators is the Iteratively Reweighted Least Squares (IRLS) method.

IRLS solves the following problem:

$$\underset{\mathbf{x}}{\text{argmin}} \sum_{i=1}^{n} \left| f_i - h_i(\mathbf{x}) \right|^p, \tag{3.30}$$

using an iterative method in which each step involves solving a weighted least squares problem of the form:

$$\mathbf{x}^{k+1} = \underset{\mathbf{x}}{\text{argmin}} \sum_{i=1}^{n} w_i \left(f_i - h_i(\mathbf{x}) \right)^2. \tag{3.31}$$

A correction factor of the weight w_i is introduced, and $(f_i - h_i(\mathbf{x}))^2$ is minimized using WLS. The initial weight factors are chosen as $w_i = 1$, and in subsequent iterations they are calculated as

$$w_i^{k+1} = w_i^k \left| f_i - h_i(\mathbf{x}) \right|^{p-2}. \tag{3.32}$$

It was shown that the IRLS method is less prone to numerical problems than the Newton–Raphson method in the case of the SHGM estimator [8]. It was also shown that IRLS almost always converges on a solution in a few iterations from the flat-start initial solution [8].

3.6   Relation to Our Network Model

The discussed numerical methods were implemented in software using a combination of the C++ and MATLAB tools [9]. Besides the solution method itself, the implementation requires an algorithm for a computation of the initial solution, first-order partial derivatives for all the measurement functions, and a configuration of measurements that makes the power-distribution system observable.

3.6.1   Determination of the Initial Solution

In our implementation, the flat-start solution is chosen as the initial solution. The solution assumes nominal voltage magnitudes (i.e., 1 per unit) and phase angles of an unloaded feeder. In the determination of the initial phase angles, phase shifts that occur due to different transformer connections need to be considered. The algorithm below presents a recursive algorithm that calculates the initial phase angles for a radial feeder.

Algorithm for the computation of initial phase angles

Data Vector of phase angles, Bus ID
Result Vector of initial phase angles

current value = angle + node phase rotation(Bus ID);
initial value(Bus ID) = current value;
ForEach *child* in *Bus ID*
      do call this method (vector of phase angles, child node ID);

Hachtel's method, as well as the Gauss–Newton and Newton–Raphson methods, requires first-order partial derivatives with respect to the measurement functions. The derivatives of the measurement functions were computed analytically, and are presented in the following subsections.

3.6.2   Partial Derivatives for Power Flow Equations

$$\frac{\partial P_{km}^a}{\partial |V_k^a|} = \sum_{j \in \Omega_p} \sum_{n \in \Omega_p} \sum_{i \in \Omega_p} \left\{ \left[(g_{km}^{jn} + g_{km,sh}^{jn}) t_{km}^{ni} t_{km}^{aj} |V_k^i| cos(\phi_k^a - \phi_k^i) \right. \right.$$

$$\left. \left. + (b_{km}^{jn} + b_{km,sh}^{jn}) t_{km}^{ni} t_{km}^{aj} |V_k^i| sin(\phi_k^a - \phi_k^i) \right] - \left[g_{km}^{jn} t_{km}^{aj} t_{km}^{ni} |V_m^i| cos(\phi_k^a - \phi_m^i) \right. \right.$$

$$
\left. - g_{km}^{jn} t_{km}^{aj} t_{km}^{ni} |V_m^i| sin(\phi_k^a - \phi_m^i) \right] \right\} + \sum_{j \in \Omega_p} \sum_{n \in \Omega_p} |V_k^a| t_{km}^{na} t_{km}^{aj} \left(g_{km}^{jn} + g_{km,sh}^{jn} \right)
$$

(3.33)

$$
\frac{\partial P_{km}^a}{\partial |V_m^i|} = -|V_k^a| \sum_{j \in \Omega_p} \sum_{n \in \Omega_p} \left[g_{km}^{jn} t_{km}^{aj} t_{mk}^{ni} cos(\phi_k^a - \phi_m^i) + b_{km}^{jn} t_{mk}^{ni} t_{km}^{aj} sin(\phi_k^a - \phi_m^i) \right]
$$

(3.34)

$$
\frac{\partial P_{km}^a}{\partial |V_k^o|} = |V_k^a| \sum_{j \in \Omega_p} \sum_{n \in \Omega_p} \left[\left(g_{km}^{jn} + g_{km,sh}^{jn} \right) t_{km}^{ni} t_{km}^{aj} cos(\phi_k^a - \phi_k^i) + \left(b_{km}^{jn} + b_{km,sh}^{jn} \right) t_{km}^{ni} \right.
$$
$$
\left. t_{km}^{aj} sin(\phi_k^a - \phi_k^i) \right]
$$

(3.35)

$$
\frac{\partial P_{km}^a}{\partial \phi_k^a} = |V_k^a| \sum_{j \in \Omega_p} \sum_{n \in \Omega_p} \sum_{i \in \Omega_p} \left\{ |V_k^i| \left[\left(g_{km}^{jn} + g_{km,sh}^{jn} \right) t_{km}^{ni} t_{km}^{aj} sin(\phi_k^a - \phi_k^i) \right. \right.
$$
$$
\left. + \left(b_{km}^{jn} + b_{km,sh}^{jn} \right) t_{km}^{ni} t_{km}^{aj} cos(\phi_k^a - \phi_k^i) \right] - |V_m^i| \left[- g_{km}^{jn} t_{mk}^{ni} t_{km}^{aj} sin(\phi_k^a - \phi_m^i) \right.
$$
$$
\left. \left. + b_{km}^{jn} t_{mk}^{ni} t_{km}^{aj} cos(\phi_k^a - \phi_m^i) \right] \right\} - |V_k^a| \sum_{j \in \Omega_p} \sum_{n \in \Omega_p} \left(b_{km}^{jn} + b_{km,sh}^{jn} \right) t_{km}^{na} t_{km}^{aj} |V_k^a|
$$

(3.36)

$$
\frac{\partial P_{km}^a}{\partial \phi_m^i} = -|V_k^a| \sum_{j \in \Omega_p} \sum_{n \in \Omega_p} \sum_{i \in \Omega_p} |V_m^i| \left[- g_{km}^{jn} t_{mk}^{ni} t_{km}^{aj} sin(\phi_k^a - \phi_m^i) - b_{km}^{jn} t_{mk}^{ni} t_{km}^{aj} \right.
$$
$$
\left. cos(\phi_k^a - \phi_m^i) \right]
$$

(3.37)

$$
\frac{\partial P_{km}^a}{\partial \phi_k^o} = |V_k^a| \sum_{j \in \Omega_p} \sum_{n \in \Omega_p} |V_k^i| \left[\left(g_{km}^{jn} + g_{km,sh}^{jn} \right) t_{km}^{ni} t_{km}^{aj} sin(\phi_k^a - \phi_k^i) - \left(b_{km}^{jn} + b_{km,sh}^{jn} \right) \right.
$$
$$
\left. t_{km}^{ni} t_{km}^{aj} cos(\phi_k^a - \phi_k^i) \right]
$$

(3.38)

$$
\frac{\partial Q_{km}^{a}}{\partial |V_{k}^{a}|} = \sum_{j\in\Omega_{p}}\sum_{n\in\Omega_{p}}\sum_{i\in\Omega_{p}}\left\{|V_{k}^{i}|\left[(g_{km}^{jn}+g_{km,sh}^{jn})t_{km}^{ni}t_{km}^{aj}sin(\phi_{k}^{a}-\phi_{k}^{i})\right.\right.
$$
$$
\left.-(b_{km}^{jn}+b_{km,sh}^{jn})t_{km}^{ni}t_{km}^{aj}cos(\phi_{k}^{a}-\phi_{k}^{i})\right]-|V_{m}^{i}|\left[g_{km}^{jn}t_{km}^{aj}t_{mk}^{ni}sin(\phi_{k}^{a}-\phi_{m}^{i})\right.
$$
$$
\left.\left.-g_{km}^{jn}t_{km}^{aj}t_{mk}^{ni}cos(\phi_{k}^{a}-\phi_{m}^{i})\right]\right\}-\sum_{j\in\Omega_{p}}\sum_{n\in\Omega_{p}}|V_{k}^{a}|t_{km}^{na}t_{km}^{aj}(b_{km}^{jn}+b_{km,sh}^{jn}) \quad (3.39)
$$

$$
\frac{\partial Q_{km}^{a}}{\partial |V_{m}^{i}|} = -|V_{k}^{a}|\sum_{j\in\Omega_{p}}\sum_{n\in\Omega_{p}}\left[g_{km}^{jn}t_{km}^{aj}t_{mk}^{ni}sin(\phi_{k}^{a}-\phi_{m}^{i})-b_{km}^{jn}t_{mk}^{ni}t_{km}^{aj}cos(\phi_{k}^{a}-\phi_{m}^{i})\right]
$$
$$
(3.40)
$$

$$
\frac{\partial Q_{km}^{a}}{\partial |V_{k}^{o}|} = |V_{k}^{a}|\sum_{j\in\Omega_{p}}\sum_{n\in\Omega_{p}}\left[(g_{km}^{jn}+g_{km,sh}^{jn})t_{km}^{ni}t_{km}^{aj}sin(\phi_{k}^{a}-\phi_{k}^{i})-(b_{km}^{jn}+b_{km,sh}^{jn})\right.
$$
$$
\left.t_{km}^{ni}t_{km}^{aj}cos(\phi_{k}^{a}-\phi_{k}^{i})\right]
$$
$$
(3.41)
$$

$$
\frac{\partial Q_{km}^{a}}{\partial \phi_{k}^{a}} = |V_{k}^{a}|\sum_{j\in\Omega_{p}}\sum_{n\in\Omega_{p}}\sum_{i\in\Omega_{p}}\left\{|V_{k}^{i}|\left[(g_{km}^{jn}+g_{km,sh}^{jn})t_{km}^{ni}t_{km}^{aj}cos(\phi_{k}^{a}-\phi_{k}^{i})\right.\right.
$$
$$
\left.+(b_{km}^{jn}+b_{km,sh}^{jn})t_{km}^{ni}t_{km}^{aj}sin(\phi_{k}^{a}-\phi_{k}^{i})\right]-|V_{m}^{i}|\left[-g_{km}^{jn}t_{mk}^{ni}t_{km}^{aj}cos(\phi_{k}^{a}-\phi_{m}^{i})\right.
$$
$$
\left.\left.+b_{km}^{jn}t_{mk}^{ni}t_{km}^{aj}sin(\phi_{k}^{a}-\phi_{m}^{i})\right]\right\}-|V_{k}^{a}|^{2}\sum_{j\in\Omega_{p}}\sum_{n\in\Omega_{p}}(g_{km}^{jn}+g_{km,sh}^{jn})t_{km}^{na}t_{km}^{aj}
$$
$$
(3.42)
$$

$$
\frac{\partial Q_{km}^{a}}{\partial \phi_{m}^{i}} = -|V_{k}^{a}|\sum_{j\in\Omega_{p}}\sum_{n\in\Omega_{p}}\sum_{i\in\Omega_{p}}|V_{m}^{i}|\left[-g_{km}^{jn}t_{mk}^{ni}t_{km}^{aj}cos(\phi_{k}^{a}-\phi_{m}^{i})+b_{km}^{jn}t_{mk}^{ni}t_{km}^{aj}\right.
$$
$$
\left.sin(\phi_{k}^{a}-\phi_{m}^{i})\right]
$$
$$
(3.43)
$$

$$
\frac{\partial Q_{km}^{a}}{\partial \phi_{k}^{o}} = |V_{k}^{a}|\sum_{j\in\Omega_{p}}\sum_{n\in\Omega_{p}}|V_{k}^{i}|\left[(g_{km}^{jn}+g_{km,sh}^{jn})t_{km}^{ni}t_{km}^{aj}cos(\phi_{k}^{a}-\phi_{k}^{i})-(b_{km}^{jn}+b_{km,sh}^{jn})\right.
$$
$$
\left.t_{km}^{ni}t_{km}^{aj}sin(\phi_{k}^{a}-\phi_{k}^{i})\right]
$$
$$
(3.44)
$$

where $i \in \{a, b, c\}$ and $o \in \{b, c\}$.

3.6.3 Partial Derivatives for Power-Injection Equations

$$\frac{\partial P_k^a}{\partial |V_k^a|} = \sum_{m \in \kappa} \sum_{p \in \Omega_p} |V_m^p| \left(G_{km}^{ap} cos(\phi_k^a - \phi_m^p) + B_{km}^{ap} sin(\phi_k^a - \phi_m^p) \right) + |V_k^a| G_{kk}^{aa}$$

$$(3.45)$$

$$\frac{\partial P_k^a}{\partial |V_m^p|} = |V_k^a| \left(G_{km}^{ap} cos(\phi_k^a - \phi_m^p) + B_{km}^{ap} sin(\phi_k^a - \phi_m^p) \right) \tag{3.46}$$

$$\frac{\partial P_k^a}{\partial \phi_k^a} = |V_k^a| \sum_{m \in \kappa} \sum_{p \in \Omega_p} |V_m^p| \left(- G_{km}^{ap} sin(\phi_k^a - \phi_m^p) + B_{km}^{ap} cos(\phi_k^a - \phi_m^p) \right)$$

$$- |V_k^a|^2 B_{kk}^{aa} \tag{3.47}$$

$$\frac{\partial P_k^a}{\partial \phi_m^p} = |V_k^a||V_m^p| \left(G_{km}^{ap} sin(\phi_k^a - \phi_m^p) - B_{km}^{ap} cos(\phi_k^a - \phi_m^p) \right) \tag{3.48}$$

$$\frac{\partial Q_k^a}{\partial |V_k^a|} = \sum_{m \in \kappa} \sum_{p \in \Omega_p} |V_m^p| \left(G_{km}^{ap} sin(\phi_k^a - \phi_m^p) - B_{km}^{ap} cos(\phi_k^a - \phi_m^p) \right) - |V_k^a| B_{kk}^{aa}$$

$$(3.49)$$

$$\frac{\partial Q_k^a}{\partial |V_m^p|} = |V_k^a| \left(G_{km}^{ap} sin(\phi_k^a - \phi_m^p) - B_{km}^{ap} cos(\phi_k^a - \phi_m^p) \right) \tag{3.50}$$

$$\frac{\partial Q_k^a}{\partial \phi_k^a} = |V_k^a| \sum_{m \in \kappa} \sum_{p \in \Omega_p} |V_m^p| \left(G_{km}^{ap} cos(\phi_k^a - \phi_m^p) + B_{km}^{ap} sin(\phi_k^a - \phi_m^p) \right) - |V_k^a|^2 G_{kk}^{aa}$$

$$(3.51)$$

$$\frac{\partial Q_k^a}{\partial \phi_m^p} = |V_k^a||V_m^p| \left(- G_{km}^{ap} cos(\phi_k^a - \phi_m^p) - B_{km}^{ap} sin(\phi_k^a - \phi_m^p) \right) \tag{3.52}$$

Where $p \in \{a, b, c\}$ if $m \neq k$, and $p \in \{b, c\}$ if $m = k$.

3.6.4 Generation of Measurement Configurations

To make the power system observable, a measurement configuration needs to be constructed. A measurement configuration includes the topological locations and measurement types for the power system in question. Several different measurement configurations were tested in simulated scenarios. The measurement configurations include combinations of one or more of the following measurement types:

- Active or reactive branch power-flow measurements,
- Active or reactive bus power-injection measurements,
- Bus-voltage phase measurements,
- Bus-voltage magnitude measurements.

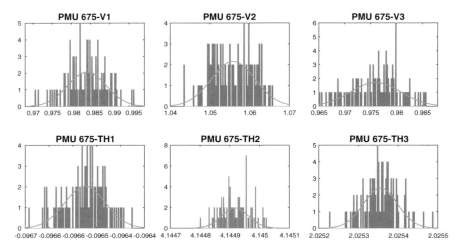

Fig. 3.2 Example of input PMU measurements

In addition to different measurement types, a particular measurement configuration can include active and reactive bus power zero injection constraints. Zero injection constraints are introduced on the buses where we are certain there is no power injection. If this information was modeled as a measurement with a correspondingly high weight (as the information is certain), it would negatively influence the conditioning of the system.

The measurements were generated from the load-flow results for the reference feeder loading. A measurement with added noise was calculated as follows:

$$z_{\text{meas}} = z_{\text{ref}} + \mathcal{N}\left(0, z_{\text{ref}} \cdot \frac{z_{\text{acc}}}{2}\right). \tag{3.53}$$

Measurement-device accuracies were modeled as 2-sigma values for the Gaussian random generator, and the values for different devices were selected as follows:

- 10% of the reference value for active and reactive power—flow measurements.
- 20% of the reference value for active and reactive power—injection measurements.
- 1% of the reference value for the PMU magnitude.
- 0.005° for the PMU angle.

An example of the input data (PMU measurements) is depicted in Fig. 3.2. For each measurement configuration, 100 Monte Carlo runs were performed.

3.7 Performance Evaluation on the Reference IEEE Power-Distribution Feeders

The implementation of numerical solvers was tested on the standard IEEE distribution test feeders [10] with 13 and 123 buses. The measurements were generated from the load-flow results for the reference feeder loading provided by the IEEE [10] (available in Appendix A). On the 13-bus feeder (depicted in Fig. 3.3), the following measurement configurations were tested

1. Power-injection measurements on every bus. No other measurements.
2. Voltage-phasor measurements on every bus. No other measurements.
3. Power-flow measurements on every branch. No other measurements.
4. The measurement configuration is listed in Table 3.1.

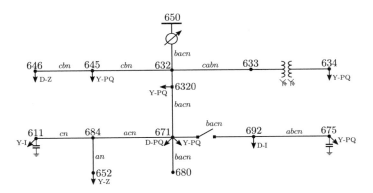

Fig. 3.3 IEEE 13-bus feeder

Table 3.1 Mixed measurement configuration for validation

Bus/branch	PQ	PMU	Flow	Zero constraint
632−633		•	•	•
633−634		•	•	•
634		•	•	
645−632		•	•	
646−645		•	•	
6320−632		•	•	
671−6320	•	•	•	
692−671	•	•	•	
675−692		•	•	
680−671		•	•	•
684−671		•	•	•
652−684		•	•	
611−684		•	•	

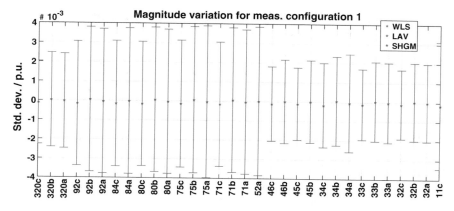

Fig. 3.4 Magnitude variation for measurement configuration 1

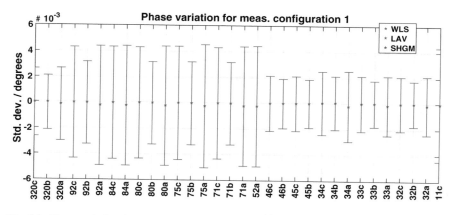

Fig. 3.5 Phase variation for measurement configuration 1

For each of the listed measurement configurations, 100 Monte Carlo (MC) simulations were generated. The MC simulations were run on all the reviewed estimators (LAV, WLS, and SHGM), where the expected values of the state estimation results were compared to the load-flow results that were used for the generation of measurements.

Figures 3.4 and 3.5 depict the voltage phase and magnitude errors (bias and standard deviation) of the SE results for the first measurement configuration and all three estimators. Please note that only last two numbers of bus/branch number are shown. The load-flow results are considered as true values. It is clear that the results among the different estimators are practically identical. It is also clear that all three estimators yield a true value, on average, which aligns with the fact that the estimators are unbiased.

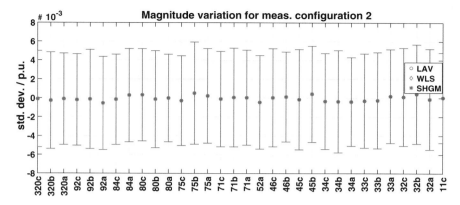

Fig. 3.6 Magnitude variation for measurement configuration 2

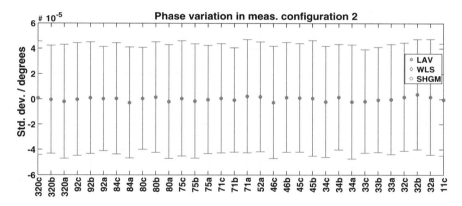

Fig. 3.7 Phase variation for measurement configuration 2

Figures 3.6 and 3.7 show practically identical behavior of the state estimators for measurement configuration two. The same can be observed also for measurement configuration three.

The fourth (mixed) measurement configuration yields more interesting results. From Figs. 3.8 and 3.9, it is clear that the WLS estimator greatly outperforms the LAV and SHGM in terms of result variance.

To observe the behavior result variance when the measurements are added to the configuration, the following experiment was conducted. The third measurement configuration (flow measurements only) was selected as the starting point, then another eight measurement configurations were created, wherein each, measurements and equality constraints were added to the previous one. Table 3.2 lists the configurations and the added measurements. For each configuration, 100 Monte Carlo runs were calculated, with the measurement accuracies described in the previous subsection. For every measurement configuration, an average standard deviation for phase and

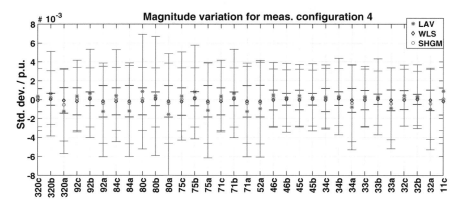

Fig. 3.8 Magnitude variation for measurement configuration 4

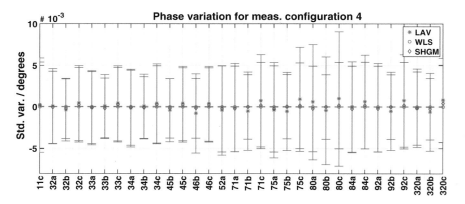

Fig. 3.9 Phase variation for measurement configuration 4

magnitude was calculated. The results of the average standard deviations for the listed measurement configurations are depicted in Figs. 3.11 and 3.10. It is clear that added measurements cause the result variance to decrease in the case of the WLS estimator. However, this is not the case with the LAV and SHGM estimators, where it is clear that the performance of the LAV and SHGM estimators is not significantly influenced by the addition of measurements. The result implies that the WLS estimator is statistically more efficient than the LAV and SHGM.

3.7.1 Computation Times for the Different Estimators

In the context of larger networks, the computational complexity of the estimator is another important aspect. In the discussed estimators, two numerical solution

Table 3.2 Measurement configurations for the IEEE 13 bus feeder

Meas. confg.	2		3	4	5	6	7	8	9
Buses	PQ	Zero const.	PMU	PMU	PMU	PMU	PMU	PMU	PMU
632		•							•
633		•							•
634								•	
645								•	
646							•		
6320							•		
671	•					•			
692	•					•			
675					•				
680		•		•					
684		•		•					
652			•						
611			•						

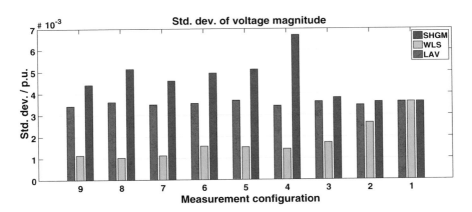

Fig. 3.10 Magnitude variation for measurement conf. listed in Table 3.2

algorithms are used, i.e., the Iteratively Reweighted Least Squares (IRLS) and Hachtel's method. Besides the algorithm and the network size, the measurement configuration (number, type, and location of measurements) is also an important factor.

Several comparisons were conducted on the IEEE 123 bus feeder (depicted in Fig. 3.12). We measured the computation times for the different estimators (WLS, SHGM, and LAV), and four different measurement configurations. The measurement configurations were as follows:

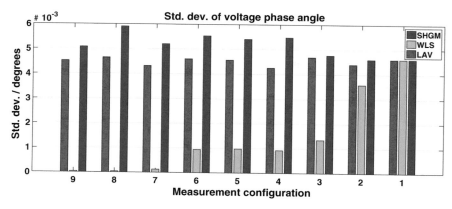

Fig. 3.11 Phase variation for measurement conf. listed in Table 3.2

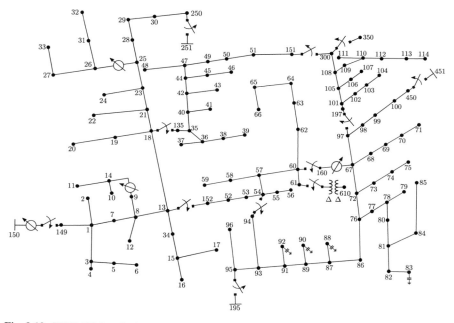

Fig. 3.12 IEEE 123-bus feeder

1. Power-injection measurements on every bus. No other measurements.
2. Voltage-phasor measurements on every bus. No other measurements.
3. Power-flow measurements on every branch. No other measurements.
4. The measurement configuration is listed in Table A.12.

The measurement results are listed in Table 3.3. It is clear that in terms of the measurement configurations, the injection measurements pose the greatest burden,

Table 3.3 Computation times

Measurement configuration/estimator	WLS	LAV	SHGM
PQ injection measurements	3.75	0.57	3.75
Voltage PMU measurements	0.12	0.17	3.00
PQ flow measurements	0.34	0.32	3.20
Configuration 1 in Table A.12	0.58	0.48	3.54

Table 3.4 Computation times

Estimator/measurement configuration	2	3	4	5	6	7
WLS	0.66	0.69	0.8	0.91	0.88	0.77
LAV	0.52	0.58	0.55	0.66	0.73	0.72
SHGM	3.67	3.69	3.82	3.98	3.83	3.67

followed by the power-flow measurements and the PMU measurements. The computational times depend on the complexity of the measurement expressions, which depend on the selection of the state variables. In terms of the different estimators, it is clear that the LAV is generally the fastest, followed by the WLS and then the SHGM. All three estimators use the Gauss–Newton solver, whereas the LAV and SHGM estimators use the IRLS algorithm on top of it for translation of the original problem into a least squares problem. It must be pointed out that since the estimation problem is in general nonlinear, the computation time does not necessarily scale with the number and type of measurements. To illustrate this fact, an experiment was conducted where the measurement configuration 1 in Table A.12 was taken as the starting configuration. Then, in each subsequent configuration, selected measurements were added to the previous measurement configuration. Table 3.4 lists the results.

References

1. A. Abur, A.G. Exposito, *Power System State Estimation: Theory and Implementation* (Marcel Dekker, New York, 2004). ISBN: 978-0-8247-5570-6
2. A. Gjelsvik, S. Aam, L. Holten, Hachtel's augmented matrix method- a rapid method improving numerical stability in power system static state estimation. IEEE Trans. Power Appar. Syst. **11**, 2987–2993 (1985)
3. A. Simoes-Costa, V.H. Quintana, An orthogonal row processing algorithm for power system sequential state estimation. IEEE Trans. Power Appar. Syst. PAS-100(8), 3791–3800 (1981). ISSN: 0018-9510, https://doi.org/10.1109/TPAS.1981.317022
4. N.D. Rao, S.C. Tripathy, Power system static state estimation by the Levenberg-Marquardt algorithm. IEEE Trans. Power Appar. Syst. (2), 695–702 (1980). Accessed 16 Aug 2017
5. I.S. Duff, J.K. Reid, A comparison of some methods for the solution of sparse overdetermined systems of linear equations. IMA J. Appl. Math. **17**(3), 267–280 (1976). Accessed 16 Aug 2017

6. A. George, M.T. Heath, Solution of sparse linear least squares problems using givens rotations. Linear Algebra Appl. **34**, 69–83 (1980)
7. D.G. Luenberger, Yinyu Ye, *Linear and Nonlinear Programming*. International Series in Operations Research and Management Science, 3rd edn. (Springer, New York, 2008). ISBN: 978-0-387-74502-2
8. L. Mili et al., Robust state estimation based on projection statistics [of power systems]. IEEE Trans. Power Syst. **11**(2), 1118–1127 (1996)
9. U. Kuhar, Three-phase state estimation in power distribution systems. PhD thesis, Jožef Stefan International Postgraduate School Ljubljana, Slovenia, 2018
10. W.H. Kersting, Radial distribution test feeders, in *Power Engineering Society Winter Meeting, 2001, IEEE*, vol. 2 (IEEE, 2001), pp. 908–912. Accessed 16 Oct 2015

Chapter 4
Small-Model and Measurement-Error Sensitivities

4.1 Introduction

Network model parameters, such as series-impedances or shunt admittances (self and mutual), that are derived from topology data (conductor geometry) can be incorrect. This is due to inaccurate data provided by the manufacturer or installation crew, inaccurate measurement campaigns, effects of aging, etc. These issues are even more prevalent in distribution networks, because the networks are larger and harder to keep track of. An uncertain data model has an influence over the estimated state variables. Thus, a method that can quickly evaluate the sensitivities, i.e., the impact of uncertain parameters on the state estimation results, can be of great benefit [1, 2].

Several methods were applied to the problem of sensitivity analyses in a power system state estimation. In [3], the authors presented a method that compares the covariance matrices of a true state vector based on the error-free model, and a state vector that is based on a model with a known error. The comparison of the estimation bias and variance gives a good measure for evaluating the modeling errors. The drawback of this method is that the complete calculation has to be repeated for every parameter that needs to be evaluated in terms of the sensitivity. This quickly becomes unfeasible as the networks become larger than a few dozen buses. In [4], the authors proposed a similar technique that allows the computation of uncertainty sensitivity coefficients for the complete measurement set. Their work, however, does not address the sensitivity calculation for the topology parameters and for the quantities that are derived from the state variables, such as power flows or power injections. In [5], the authors proposed a two-step method for uncertainty analysis. First, a weighted least squares state estimation solution is computed and then the upper and lower interval bounds are obtained using linear programming. If the inaccuracy in the measurements is modeled by a probability distribution function, then the calculated state vector is also modeled by a probability distribution function. Since the statistics of the measurement errors are difficult to characterize in practice, the idea here is that the accuracy of a particular measurement is described as an interval, for example $\pm 2\%$,

© The Author(s), under exclusive license to Springer Nature Switzerland AG 2020
U. Kuhar et al., *Observability of Power-Distribution Systems*,
SpringerBriefs in Applied Sciences and Technology,
https://doi.org/10.1007/978-3-030-39476-9_4

rather than by the standard deviation or variance. The second step of the analysis allows the specification of not just a single optimal estimate of the particular state variable, but also an uncertainty range within which one can be certain that the true value lies. The limitation of the method is that the lower and upper bounds need to be computed separately for each state variable and for each measurement. In [6], the authors expanded this approach to also include the network parameters. In [7], a similar approach using an interval-analysis technique was proposed to find the bounds of the state variables of the power system whose line parameters lie within particular bounds. A linear measurement model is assumed (i.e., voltage-phasor measurements only), other measurements are included by using the equivalent measurement technique (i.e., measurements are converted into their equivalent voltage-phasor values). In [8, 9], the authors developed analytical expressions for the propagation of uncertainty of phasor measurement units (PMUs) in the state estimation algorithm. In [10], the authors also describe the use of the Monte Carlo method for the evaluation of different uncertainty sources on a three-phase state estimator. In [11], Mínguez and Conejo leveraged the perturbation approach to sensitivity analysis based on the differentiation of the Karush–Kuhn–Tucker (KKT) conditions, described in [12]. We generalized their work with a full three-phase analysis and with the expression of uncertainties in terms of intervals. Since the calculated sensitivities are only valid at the optimal solution point, we also suggest an upper error bound on the interval analysis.

4.2 Perturbation Approach to Sensitivity Analyses

A sensitivity analysis consists of determining how and how much a certain parameter change in the optimization problem influences the objective function value. The existing methods for sensitivity analysis share the following limitations [12]:

- There are different methods for different cases of obtaining sensitivities, but no integrated approach that yields all the sensitivities at once.
- They assume the existence of the partial derivatives of the objective function with respect to the parameters, which is not always the case.
- They assume that the active constraints remain active, which implies that there is no need to distinguish between equality and inequality constraints.

With the perturbation approach, a general analysis can be performed without assuming the existence of the partial derivatives and without assuming that the active inequality constraints remain active. Also, the perturbation approach yields all the sensitivities with respect to the parameters at once.

Considering the cost function, the state estimation problem can be formulated as an optimization problem including constraints:

$$\underset{\mathbf{x}}{\text{minimize}} \quad J(\mathbf{x}, \mathbf{a})$$

$$\text{subject to} \quad \mathbf{c}(\mathbf{x}, \mathbf{a}) = 0 : \boldsymbol{\lambda}, \quad\quad (4.1)$$

$$\mathbf{g}(\mathbf{x}, \mathbf{a}) \leq 0 : \boldsymbol{\mu},$$

where $\mathbf{x} \in \mathbb{R}^n$, $\mathbf{a} \in \mathbb{R}^p$, $J(\mathbf{x}, \mathbf{a})$ is a scalar cost function, $\mathbf{c}(\mathbf{x}, \mathbf{a})$ is the vector of equality constraints representing exactly known pseudo-measurements (zero injections), $\mathbf{g}(\mathbf{x}, \mathbf{a})$ is the vector of inequality constraints (e.g., the phase angles must be smaller than $\frac{\pi}{2}$), and $\boldsymbol{\lambda}$ and $\boldsymbol{\mu}$ are the Lagrange multiplier vectors for the equality and inequality constraints, respectively. Note that vector $\mathbf{a}$ includes measurements and model data that are of interest in the calculation of sensitivities. In our case, it includes all the measurements and all the system-admittance elements, i.e., $\mathbf{G}$, $\mathbf{B}$, $\mathbf{b}_{shunt}$, and $\mathbf{g}_{shunt}$.

4.2.1 Dual Variables

Duality is a crucial concept in the following sensitivity analysis of a nonlinear optimization problem. Consider the nonlinear *primal* problem in Eq. 4.1. The *dual* problem requires the introduction of the dual function, defined as

$$\Theta(\boldsymbol{\lambda}, \boldsymbol{\mu}) = \underset{\mathbf{x}}{\text{infimum}} \left[J(\mathbf{x}, \mathbf{a}) + \boldsymbol{\lambda}^T \mathbf{c}(\mathbf{x}, \mathbf{a}) + \boldsymbol{\mu}^T \mathbf{g}(\mathbf{x}, \mathbf{a}) \right] \quad\quad (4.2)$$

The *dual* problem is then defined as follows:

$$\underset{\mathbf{x}}{\text{maximize}} \quad J(\mathbf{x}, \mathbf{a}),$$

$$\text{subject to} \quad \boldsymbol{\mu} \geq 0 \quad\quad (4.3)$$

Using the Lagrangian:

$$\mathscr{L}(\mathbf{x}, \boldsymbol{\lambda}, \boldsymbol{\mu}) = J(\mathbf{x}, \mathbf{a}) + \boldsymbol{\lambda}^T \mathbf{c}(\mathbf{x}, \mathbf{a}) + \boldsymbol{\mu}^T \mathbf{g}(\mathbf{x}, \mathbf{a}), \quad\quad (4.4)$$

we can rewrite the *dual problem* as

$$\underset{\boldsymbol{\lambda}, \boldsymbol{\mu}; \boldsymbol{\mu} \geq 0}{\text{maximize}} \left(\underset{\mathbf{x}}{\text{infimum}} \mathscr{L}(\mathbf{x}, \boldsymbol{\lambda}, \boldsymbol{\mu}) \right) \quad\quad (4.5)$$

It is assumed that J, $\mathbf{c}$, and $\mathbf{g}$ are such that the infimum of the Lagrangian function is always attained at some $\mathbf{x}$, so that the infimum operation in Eqs. 4.2 and 4.5 can be replaced by the minimum operation. Then, the problem in Eq. 4.5 is referred to as the *max-min* dual problem [13].

4.2.2 Sensitivities

The sensitivities around the current operating point are of interest. As such, the first step toward obtaining the sensitivities is to calculate the optimal solution for the current state estimation problem, i.e., the problem in Eq. 4.1 needs to be solved. The solution is denoted as $(\mathbf{x}^*, \mathbf{a}, \boldsymbol{\lambda}^*)$. Note that inactive, i.e., nonbinding, inequality constraints are disregarded, and the active ones are incorporated as equality constraints. Then, the KKT conditions are stated for the optimal solution:

$$\nabla_{\mathbf{x}} J(\mathbf{x}^*, \mathbf{a}) + \sum_{k=1}^{o} \lambda_k^* \nabla_{\mathbf{x}} c_k(\mathbf{x}^*, \mathbf{a}) + \sum_{j=1}^{l} \mu_j^* \nabla_{\mathbf{x}} g_j(\mathbf{x}^*, \mathbf{a}) = \mathbf{0} \tag{4.6}$$

$$c_k(\mathbf{x}^*, \mathbf{a}) = 0, k = 1, 2, \ldots, o, \tag{4.7}$$

$$g_j(\mathbf{x}^*, \mathbf{a}) \le 0, j = 1, 2, \ldots, l \tag{4.8}$$

$$\mu_j^* g_j(\mathbf{x}^*, \mathbf{a}) \le 0, j = 1, 2, \ldots, l \tag{4.9}$$

$$\mu_j^* \ge 0, j = 1, 2, \ldots, l \tag{4.10}$$

where $\nabla_{\mathbf{x}}$ represents the gradient with respect to vector $\mathbf{x}$, and l and r are the numbers of the equality and inequality constraints, respectively. The conditions in Eqs. 4.7–4.8 are the primal feasibility conditions, the conditions in Eq. 4.9 are the complementary slackness conditions, and the conditions in Eq. 4.10 impose the non-negativity of the multipliers of the inequality constraints and are referred to as the dual feasibility conditions.

As the sensitivity analysis needs information about the dual problem, the required step before the sensitivity study is to obtain the dual variables solving the linear system of equations that is obtained from Eqs. 4.6–4.10:

$$\begin{bmatrix} \frac{\partial J}{\partial x_1} \\ \frac{\partial J}{\partial x_2} \\ \vdots \\ \frac{\partial J}{\partial x_n} \end{bmatrix} = \begin{bmatrix} \frac{\partial c_1}{\partial x_1} & \frac{\partial c_2}{\partial x_1} & \cdots & \frac{\partial c_o}{\partial x_1} \\ \frac{\partial c_1}{\partial x_2} & \frac{\partial c_2}{\partial x_2} & \cdots & \frac{\partial c_o}{\partial x_2} \\ \vdots & \vdots & \vdots & \vdots \\ \frac{\partial c_1}{\partial x_n} & \frac{\partial c_2}{\partial x_n} & \cdots & \frac{\partial c_o}{\partial x_n} \end{bmatrix} \cdot \begin{bmatrix} \lambda_1 \\ \lambda_2 \\ \vdots \\ \lambda_o \end{bmatrix}. \tag{4.11}$$

If the gradient vectors of the binding constraints at a solution are linearly independent, the solution of the obtained system is guaranteed to be unique.

In the next step, the set of variables $\mathbf{x}^*, \mathbf{a}, \boldsymbol{\lambda}^*, \boldsymbol{\mu}^*, J^*$ is perturbed in a way that the KKT conditions still hold [12]. Thus, to obtain the sensitivity equations, Eq. 4.1 and 4.6–4.10 are differentiated as follows:

$$\left[\nabla_{\mathbf{x}} J(\mathbf{x}^*, \mathbf{a}) \right]^T d\mathbf{x} + \left[\nabla_{\mathbf{a}} J(\mathbf{x}^*, \mathbf{a}) \right]^T d\mathbf{a} - dJ = 0 \tag{4.12}$$

$$\left[\nabla_{\mathbf{xx}} J(\mathbf{x}^*, \mathbf{a}) + \sum_{k=1}^{o} \lambda_k^* \nabla_{\mathbf{xx}} c_k(\mathbf{x}^*, \mathbf{a}) + \sum_{j=1}^{l} \mu_j^* \nabla_{\mathbf{xx}} g_j(\mathbf{x}^*, \mathbf{a}) \right] d\mathbf{x}$$

$$+ \left[\nabla_{\mathbf{xa}} J(\mathbf{x}^*, \mathbf{a}) + \sum_{k=1}^{o} \lambda_k^* \nabla_{\mathbf{xa}} c_k(\mathbf{x}^*, \mathbf{a}) + \sum_{j=1}^{l} \mu_j^* \nabla_{\mathbf{xa}} g_j(\mathbf{x}^*, \mathbf{a}) \right] d\mathbf{a}$$

$$+ \nabla_{\mathbf{x}} \mathbf{c}(\mathbf{x}^*, \mathbf{a}) d\lambda + \nabla_{\mathbf{x}} \mathbf{g}(\mathbf{x}^*, \mathbf{a}) d\mu = 0 \tag{4.13}$$

$$\left[\nabla_{\mathbf{x}} \mathbf{c}(\mathbf{x}^*, \mathbf{a}) \right]^T d\mathbf{x} + \left[\nabla_{\mathbf{a}} \mathbf{c}(\mathbf{x}^*, \mathbf{a}) \right]^T d\mathbf{a} = 0 \tag{4.14}$$

$$\left[\nabla_{\mathbf{x}} g_j(\mathbf{x}^*, \mathbf{a}) \right]^T d\mathbf{x} + \left[\nabla_{\mathbf{a}} g_j(\mathbf{x}^*, \mathbf{a}) \right]^T d\mathbf{a} = 0$$
$$\text{if } \mu_j^* \neq 0, \, j \in \Omega_I \tag{4.15}$$

where $\nabla_{\mathbf{a}}$ represents the gradient with respect to vector $\mathbf{a}$, $\nabla_{\mathbf{xx}}$ and $\nabla_{\mathbf{xa}}$ represent the Hessian with respect to the $\mathbf{x}, \mathbf{x}$ and $\mathbf{x}, \mathbf{a}$ vectors, respectively. Ω_I is a set of active inequality constraints, and all the matrices are evaluated at the optimal solution $\mathbf{x}^*, \lambda^*$, μ^*, J^*. For the solution of the estimation problem, the binding inequality constraints are considered as equality constraints, and nonbinding ones are disregarded. It turns out that the inequality constraints generally do not have an influence on the solution of the state estimation problems, since the equality constraints tightly condition that solution.

The system of equations is then formed as follows:

$$\begin{bmatrix} \mathbf{F_x} & \mathbf{F_a} & \mathbf{0} & -1 \\ \mathbf{F_{xx}} & \mathbf{F_{xa}} & \mathbf{C_x}^T & 0 \\ \mathbf{C_x} & \mathbf{C_a} & \mathbf{0} & 0 \end{bmatrix} \cdot \begin{bmatrix} d\mathbf{x} \\ d\mathbf{a} \\ d\lambda \\ dJ \end{bmatrix} = \mathbf{0} \tag{4.16}$$

where the vectors and submatrices (including their dimensions) are

$$\mathbf{F_{x(1 \times n)}} = \left[\nabla_{\mathbf{x}} J(\mathbf{x}^*, \mathbf{a}) \right]^T, \tag{4.17}$$

$$\mathbf{F_{a(1 \times p)}} = \left[\nabla_{\mathbf{a}} J(\mathbf{x}^*, \mathbf{a}) \right]^T, \tag{4.18}$$

$$\mathbf{F_{xx(n \times n)}} = \nabla_{\mathbf{xx}} J(\mathbf{x}^*, \mathbf{a}) + \sum_{k=1}^{o} \lambda_k^* \nabla_{\mathbf{xx}} c_k(\mathbf{x}^*, \mathbf{a}), \tag{4.19}$$

$$\mathbf{F_{xa(n \times p)}} = \nabla_{\mathbf{xa}} J(\mathbf{x}^*, \mathbf{a}) + \sum_{k=1}^{o} \lambda_k^* \nabla_{\mathbf{xa}} c_k(\mathbf{x}^*, \mathbf{a}), \tag{4.20}$$

$$\mathbf{C_{x(o \times n)}} = \left[\nabla_{\mathbf{x}} \mathbf{c}(\mathbf{x}^*, \mathbf{a}) \right]^T, \tag{4.21}$$

$$\mathbf{C_{a(o \times p)}} = \left[\nabla_{\mathbf{a}} \mathbf{c}(\mathbf{x}^*, \mathbf{a}) \right]^T, \tag{4.22}$$

where p is the dimension of vector $\mathbf{a}$.

To calculate the sensitivities of the state vector with respect to the data vector, the system in Eq. 4.16 is reformulated as

$$\mathbf{U} \left[d\mathbf{x} \, d\lambda \, dJ \right]^T = \mathbf{S} d\mathbf{a} \tag{4.23}$$

Replacing $d\mathbf{a}$ by the p-dimensional identity matrix $\mathbf{I}$, the matrices $\mathbf{U}$ and $\mathbf{S}$ become

$$\mathbf{U} = \begin{bmatrix} \mathbf{F_x} & \mathbf{0} & -\mathbf{1} \\ \mathbf{F_{xx}} & \mathbf{C_x^T} & \mathbf{0} \\ \mathbf{C_x} & \mathbf{0} & \mathbf{0} \end{bmatrix}$$

$$\mathbf{S}^T = -\begin{bmatrix} \mathbf{F_a} & \mathbf{F_{xa}} & \mathbf{C_a} \end{bmatrix}$$

(4.24)

from which it follows that

$$\begin{bmatrix} \frac{\partial \mathbf{x}}{\partial \mathbf{a}} & \frac{\partial \boldsymbol{\lambda}}{\partial \mathbf{a}} & \frac{\partial J}{\partial \mathbf{a}} \end{bmatrix}^T = \mathbf{U}^{-1}\mathbf{S}.$$

(4.25)

In Eq. 4.25 the sensitivities of the state variables, Lagrange multipliers, and the objective function are expressed with respect to the data vector. Equation 4.25 can be solved as follows:

$$\begin{bmatrix} \frac{\partial \mathbf{x}}{\partial \mathbf{a}} & \frac{\partial \boldsymbol{\lambda}}{\partial \mathbf{a}} \end{bmatrix}^T = -\mathbf{H_x}^{-1}\mathbf{H_a}$$

(4.26)

$$\frac{\partial J}{\partial \mathbf{a}} = \mathbf{F_a} + \mathbf{F_x}\frac{\partial \mathbf{x}}{\partial \mathbf{a}}$$

(4.27)

where the matrices $\mathbf{H}_x$ and $\mathbf{H}_a$ are

$$\mathbf{H}_x = \begin{bmatrix} \mathbf{F_{xx}} & \mathbf{C_x}^T \\ \mathbf{C_x} & \mathbf{0} \end{bmatrix}$$

$$\mathbf{H}_a = \begin{bmatrix} \mathbf{F_{xa}} \\ \mathbf{C_a} \end{bmatrix}.$$

(4.28)

If the power system under consideration is observable, then the matrices $\mathbf{U}$ and $\mathbf{H}_x$ are invertible. Since distribution networks have radial or very weakly meshed topologies, the matrices $\mathbf{U}$ and $\mathbf{H}_x$ are very sparse, and so they are easily factorized using sparse-oriented algorithms.

The solution to the system in Eq. 4.25 yields sensitivity expressions with respect to the data vector $\mathbf{a}$, which contains measurements and system-admittance elements. From an engineering perspective, the important parameters are the sensitivities with respect to the conductor lengths, i.e., their impedances (since the conductor resistances and reactances are linearly dependent on the conductor's length [14]). The chain rule must be used to obtain the sensitivities with respect to the line impedances from sensitivities with respect to the admittance elements. The partial derivatives for a state variable with respect to the conductor resistance and reactance are

$$\frac{\partial x}{\partial r_{ij}^{mn}} = \frac{\partial G_{ij}^{mn}}{\partial r_{ij}^{mn}}\left(\frac{\partial x}{\partial G_{ij}^{mn}} + \frac{\partial x}{\partial G_{ji}^{mn}} - \frac{\partial x}{\partial G_{ii}^{mn}} - \frac{\partial x}{\partial G_{jj}^{mn}}\right)$$

$$+ \frac{\partial B_{ij}^{mn}}{\partial r_{ij}^{mn}}\left(\frac{\partial x}{\partial B_{ij}^{mn}} + \frac{\partial x}{\partial B_{ji}^{mn}} - \frac{\partial x}{\partial B_{ii}^{mn}} - \frac{\partial x}{\partial B_{jj}^{mn}}\right)$$

(4.29)

$$\frac{\partial x}{\partial x_{ij}^{mn}} = \frac{\partial G_{ij}^{mn}}{\partial x_{ij}^{mn}}\left(\frac{\partial x}{\partial G_{ij}^{mn}} + \frac{\partial x}{\partial G_{ji}^{mn}} - \frac{\partial x}{\partial G_{ii}^{mn}} - \frac{\partial x}{\partial G_{jj}^{mn}}\right)$$
$$+ \frac{\partial B_{ij}^{mn}}{\partial x_{ij}^{mn}}\left(\frac{\partial x}{\partial B_{ij}^{mn}} + \frac{\partial x}{\partial B_{ji}^{mn}} - \frac{\partial x}{\partial B_{ii}^{mn}} - \frac{\partial x}{\partial B_{jj}^{mn}}\right), \tag{4.30}$$

where partial derivatives with respect to the system-admittance elements are known from Eq. 4.25, and the admittance element partial derivatives are

$$\frac{\partial G_{ij}^{mn}}{\partial r_{ij}^{mn}} = \frac{\partial B_{ij}^{mn}}{\partial x_{ij}^{mn}} = \frac{(r_{ij}^{mn})^2 - (x_{ij}^{mn})^2}{\left((r_{ij}^{mn})^2 + (x_{ij}^{mn})^2\right)^2} \tag{4.31}$$

$$\frac{\partial G_{ij}^{mn}}{\partial x_{ij}^{mn}} = -\frac{\partial B_{ij}^{mn}}{\partial r_{ij}^{mn}} = \frac{2 r_{ij}^{mn} x_{ij}^{mn}}{\left((r_{ij}^{mn})^2 + (x_{ij}^{mn})^2\right)^2}. \tag{4.32}$$

Besides the sensitivities of the state variables, the sensitivities of the derived variables, such as power flows or power injections, are also of interest. The way to obtain them is, again, by applying the chain rule. For example, the sensitivity of the power flow in phase a with respect to its self-series resistance is

$$\frac{\partial P_{km}^a}{\partial g_{km}^{aa}} = \frac{\partial P_{km}^a}{\partial g_{km}^{aa}} + \frac{\partial P_{km}^a}{\partial |V_k^a|}\frac{\partial |V_k^a|}{\partial g_{km}^{aa}} + \frac{\partial P_{km}^a}{\partial |V_k^b|}\frac{\partial |V_k^b|}{\partial g_{km}^{aa}}$$
$$+ \frac{\partial P_{km}^a}{\partial |V_k^c|}\frac{\partial |V_k^c|}{\partial g_{km}^{aa}} + \frac{\partial P_{km}^a}{\partial |V_m^a|}\frac{\partial |V_m^a|}{\partial g_{km}^{aa}} + \frac{\partial P_{km}^a}{\partial |V_m^b|}\frac{\partial |V_m^b|}{\partial g_{km}^{aa}}$$
$$+ \frac{\partial P_{km}^a}{\partial |V_m^c|}\frac{\partial |V_m^c|}{\partial g_{km}^{aa}} + \frac{\partial P_{km}^a}{\partial \phi_k^a}\frac{\partial \phi_k^a}{\partial g_{km}^{aa}} + \frac{\partial P_{km}^a}{\partial \phi_k^b}\frac{\partial \phi_k^b}{\partial g_{km}^{aa}}$$
$$+ \frac{\partial P_{km}^a}{\partial \phi_k^c}\frac{\partial \phi_k^c}{\partial g_{km}^{aa}} + \frac{\partial P_{km}^a}{\partial \phi_m^a}\frac{\partial \phi_m^a}{\partial g_{km}^{aa}} + \frac{\partial P_{km}^a}{\partial \phi_m^b}\frac{\partial \phi_m^b}{\partial g_{km}^{aa}}$$
$$+ \frac{\partial P_{km}^a}{\partial \phi_m^c}\frac{\partial \phi_m^c}{\partial g_{km}^{aa}}. \tag{4.33}$$

4.2.3 Implementation

The perturbation approach was implemented in a computer program using the C++ language. All the derivatives were implemented numerically. The LAV and SHGM estimators produce objective functions that have discontinuous derivatives. Since we are interested in small deviations of the model and measurement parameters, we do not operate in the area of discontinuities with the SHGM estimator. This is not the case for the LAV estimator, where discontinuity is encountered at $r_i = 0$. For the LAV estimator, the problem is restated as

$$J(\mathbf{x}, \mathbf{a}) = \sum_{i=1}^{m} r_i \tag{4.34}$$

and the measurements are imposed as constraints [15], i.e., $(f_i - h_i(\mathbf{x}))/\sigma_i \leq r_i$ and $(f_i - h_i(\mathbf{x}))/\sigma_i \geq -r_i$.

For the numerical implementation, the following approximations were used

$$\frac{\partial f(a, b)}{\partial x} \approx \frac{f(a + h, b) - f(a, b)}{h} \tag{4.35}$$

$$\frac{\partial^2 f(a, b)}{\partial^2 x} \approx \frac{-f(a - h, b) + 2f(a, b) - f(a + h, b)}{h^2} \tag{4.36}$$

$$\frac{\partial^2 f(a, b)}{\partial x \partial y} \approx \frac{f(a + h_1, b + h_2) - f(a + h_1, b - h_2) - f(a - h_1, b + h_2) + f(a - h_1, b - h_2)}{4h_1 h_2}$$
$$\tag{4.37}$$

A formula for h that balances the rounding error against the secant error for optimum accuracy is

$$h = 2\sqrt{\epsilon \left| \frac{f(x)}{f''(x)} \right|} \tag{4.38}$$

4.2.3.1 Implementation Validation

The perturbation approach to the implementation of the sensitivity analysis was validated by comparing the obtained results with the direct numerical derivatives. The measurements were generated from the load-flow results for the reference feeder loading provided by the IEEE [16], more precisely the validation was performed on the 13-bus IEEE distribution test feeder [16] shown in Fig. A.1.

The validation was made on a single configuration of measurements, which is listed in Table 4.1. The PMU column indicates the presence of a PMU measurement on a corresponding bus, the PQ injection bus indicates the presence of an active/reactive power-injection measurement on the corresponding bus, and the equality constraints column indicates whether zero injection constraints are imposed on the corresponding bus. For validation purposes, the measurements were assumed to be perfectly accurate without added noise [17]. The direct numerical derivatives were computed as follows: first, the state estimation was computed to obtain the reference state vector $\mathbf{x}_{ref}$, then the parameter in question was perturbed $a + \Delta a$ and the state estimation was computed again with a perturbation to obtain the state vector $\mathbf{x}_p$, and the derivatives were then obtained as follows:

$$\frac{\partial \mathbf{x}}{\partial a} \approx \frac{\mathbf{x}_p - \mathbf{x}_{ref}}{\Delta a}. \tag{4.39}$$

Table 4.1 Measurement configuration for validation

Bus ID	PMU	PQ injection	Equality constraints
650			
632	●		●
633	●		●
634		●	
645	●		
646		●	
6320	●		
671	●		
692	●		
675		●	
680		●	●
684	●		●
652		●	
611		●	

The cases presented here are the phase-angle sensitivity across all buses with respect to the active-power measurement on bus 634, and the magnitude sensitivity across all the buses with respect to the admittance matrix element $\mathbf{Y}_{652a-684a}$. The results are shown in Figs. 4.1 and 4.2. Please note that only last two numbers of bus/branch number are shown. The lower graphs show the error with respect to the direct numerical derivative. It is clear that there is some difference between the perturbation approach results and the direct numerical derivatives. This difference is attributed to numerical errors since, in the perturbation approach, all the derivatives are computed numerically.

4.2.4 Error Bound

The sensitivities obtained with the solution to the system in Eq. 4.25 represent tangent lines at the optimal solution point. These derivatives are only truly valid for that point; however, from an engineering perspective, we are more interested in the intervals, e.g., we would like to know within which interval the state vector can be expected if we are confident that the line length is within a certain range.

The interval $[-\mathbf{x}_a, \mathbf{x}_a]$ can be obtained from

$$\pm \mathbf{x}_a \approx \mathbf{x}^* \pm \Delta\mathbf{x} = \mathbf{x}^* \pm \Delta a \frac{\partial \mathbf{x}}{\partial a}. \tag{4.40}$$

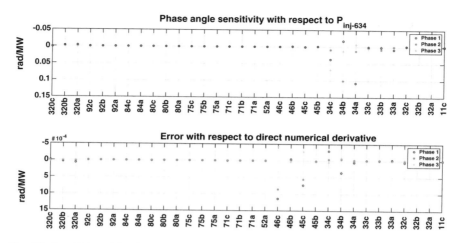

Fig. 4.1 Sensitivity of the phase angles with respect to the active-power measurement on bus 634

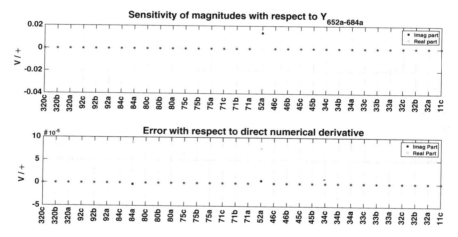

Fig. 4.2 Sensitivity of the voltage magnitudes with respect to the admittance matrix element

Since the derivative is only valid for the optimal solution, an error will invariably be made with such an interval calculation.

An example for the phase part of the state vector error is shown in Fig. 4.3, for $684 - 652$ conductor impedance ($z_{684a-652a}$). The error is computed as

$$E = max \left| \mathbf{x}_{true} - \left(\mathbf{x}^* + \Delta z_{684a-652a} \frac{\partial \mathbf{x}^*}{\partial z_{684a-652a}} \right) \right| \qquad (4.41)$$

The conductor has a specified length of 800 ft, while the computation was performed for an impedance interval corresponding to a conductor length in the range from 600

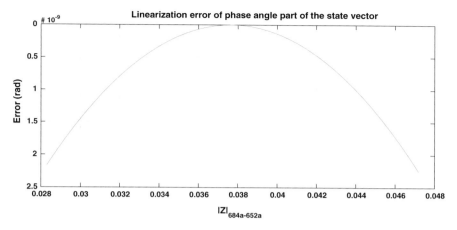

Fig. 4.3 Linearization error of the phase-angle part of the state vector

to $1000\,\text{ft}$ ($\pm 25\%$). It is clear that although the error is small, it quickly increases as we move away from the value at the optimal solution $\mathbf{x}^*$. Thus, a bound that would specify the maximum state vector error when a parameter interval analysis is performed is of interest. The error bound for the linear approximation would normally require knowledge of the second-order derivatives $\frac{\partial^2 \mathbf{x}}{\partial \mathbf{a}^2}$. Calculating them is impracticable for the problem at hand.

Another approach is to take advantage of the nature of the problem and the radial topology of distribution systems. The only two things that influence the state variables are the line impedances and the feeder loading. In radial networks, the largest decrease of the state variables (voltage magnitudes and phase angles) is achieved when the estimated feeder loading is the largest and when the given line impedances are the greatest (and vice versa). To illustrate this, simulations were performed for six different feeder loadings and line impedances. For each run, the loads as well as the line impedances were increased. Figures 4.4 and 4.5 show that the voltage magnitudes and the voltage phases decrease on all the buses for each consecutive run.

From this the rules for the data vectors $\mathbf{a}_u$ and $\mathbf{a}_l$ that will yield the state vectors with the largest deviation from $\mathbf{x}^*$ in both directions (i.e., $\mathbf{x}_u$ and $\mathbf{x}_l$) can be inferred. For power-injection and flow measurements, the upper and lower interval bounds are taken for $\mathbf{a}_u$ and $\mathbf{a}_l$, respectively:

$$\{\mathbf{P}_{i,u}, \mathbf{Q}_{i,u}, \mathbf{P}_{ij,u}, \mathbf{Q}_{ij,u}\} = \{\mathbf{P}_i, \mathbf{Q}_i, \mathbf{P}_{ij}, \mathbf{Q}_{ij}\}$$
$$+\{\mathbf{\Delta P}_i, \mathbf{\Delta Q}_i, \mathbf{\Delta P}_{ij}, \mathbf{\Delta Q}_{ij}\} \tag{4.42}$$

$$\{\mathbf{P}_{i,l}, \mathbf{Q}_{i,l}, \mathbf{P}_{ij,l}, \mathbf{Q}_{ij,l}\} = \{\mathbf{P}_i, \mathbf{Q}_i, \mathbf{P}_{ij}, \mathbf{Q}_{ij}\}$$
$$-\{\mathbf{\Delta P}_i, \mathbf{\Delta Q}_i, \mathbf{\Delta P}_{ij}, \mathbf{\Delta Q}_{ij}\} \tag{4.43}$$

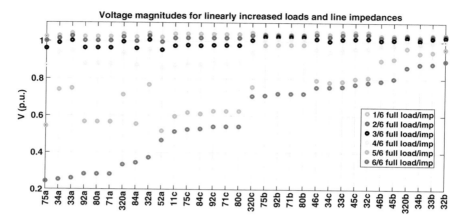

Fig. 4.4 Voltage magnitudes for linearly increased loads and impedances

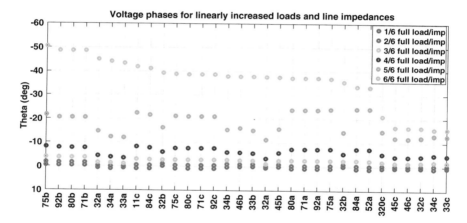

Fig. 4.5 Voltage phases for linearly increased loads and impedances

For the line resistances and reactances, the upper interval bounds are also used

$$\{\mathbf{r}_{ij,u}, \mathbf{x}_{ij,u}\} = \{\mathbf{r}_{ij}, \mathbf{x}_{ij}\} + \{\boldsymbol{\Delta}\mathbf{x}_{ij}, \boldsymbol{\Delta}\mathbf{r}_{ij}\} \tag{4.44}$$

$$\{\mathbf{r}_{ij,l}, \mathbf{x}_{ij,l}\} = \{\mathbf{r}_{ij}, \mathbf{x}_{ij}\} - \{\boldsymbol{\Delta}\mathbf{x}_{ij}, \boldsymbol{\Delta}\mathbf{r}_{ij}\} \tag{4.45}$$

For the voltage magnitudes, the lower interval bounds are used

$$|\mathbf{V}|_{i,u} = |\mathbf{V}|_i - \boldsymbol{\Delta}|\mathbf{V}|_i \tag{4.46}$$

$$|\mathbf{V}|_{i,l} = |\mathbf{V}|_i + \boldsymbol{\Delta}|\mathbf{V}|_i \tag{4.47}$$

For the voltage phase angles, the lower interval bounds are also used

$$\boldsymbol{\Theta}_{i,u} = \boldsymbol{\Theta}_i - \boldsymbol{\Delta\Theta}_i \tag{4.48}$$

$$\boldsymbol{\Theta}_{i,l} = \boldsymbol{\Theta}_i + \boldsymbol{\Delta\Theta}_i \tag{4.49}$$

The data vectors are, as a result, formed as

$$\mathbf{a}_u = \big[\mathbf{P}_{i,u}, \mathbf{Q}_{i,u}, \mathbf{P}_{ij,u}, \mathbf{Q}_{ij,u}, \mathbf{r}_{ij,u},$$
$$\mathbf{x}_{ij,u}, |\mathbf{V}|_{i,u}, \boldsymbol{\Theta}_{i,u} \big]. \tag{4.50}$$

$$\mathbf{a}_l = \big[\mathbf{P}_{i,l}, \mathbf{Q}_{i,l}, \mathbf{P}_{ij,l}, \mathbf{Q}_{ij,l}, \mathbf{r}_{ij,l},$$
$$\mathbf{x}_{ij,l}, |\mathbf{V}|_{i,l}, \boldsymbol{\Theta}_{i,l} \big]. \tag{4.51}$$

Using them, the state vectors $\mathbf{x}_u$ and $\mathbf{x}_l$ are obtained. With these obtained state vectors, the upper error bounds on the interval calculations can be specified as

$$\left| \mathbf{x}_u - \left(\mathbf{x}^* + \Delta a \frac{\partial \mathbf{x}^*}{\partial a} \right) \right| \geqslant \left| \mathbf{x}_{true} - \left(\mathbf{x}^* + \Delta a \frac{\partial \mathbf{x}^*}{\partial a} \right) \right| \tag{4.52}$$
$$\forall a \in [\mathbf{a}, \mathbf{a}_u]$$

$$\left| \mathbf{x}_l - \left(\mathbf{x}^* + \Delta a \frac{\partial \mathbf{x}^*}{\partial a} \right) \right| \geqslant \left| \mathbf{x}_{true} - \left(\mathbf{x}^* + \Delta a \frac{\partial \mathbf{x}^*}{\partial a} \right) \right| \tag{4.53}$$
$$\forall a \in [\mathbf{a}, \mathbf{a}_l]$$

These error bounds only apply to radial networks.

4.2.5 Results

The sensitivity calculations were performed for the three different measurement configurations that are listed in Table 4.2, and for the three different estimators, i.e., WLS, LAV, and SHGM. For each configuration, the sensitivities of the state variables with respect to the measurements and the conductor impedances (self and mutual) were of interest. Different measurement configurations and estimators yield a large number of possible combinations, and although we conducted an exhaustive number of simulations, we chose to depict the most representative results. The following subsections provide the analysis, simulations, and results for the uncertain conductor length, and the uncertain measurements.

Table 4.2 Measurement configuration for calculation of sensitivities

Configuration	PQ		PQ-PMU		PMU-PQ	
Bus ID	PQ	PMU	PQ	PMU	PQ	PMU
632	•		•	•	•	•
633	•		•	•	•	•
634	•		•	•		•
645	•		•	•	•	•
646	•		•	•		•
6320	•		•	•		•
671	•		•	•		•
692	•		•	•		•
675	•		•	•	•	•
680	•		•	•	•	•
684	•		•	•		•
652	•		•	•	•	•
611	•		•	•	•	•

4.2.6   Variation of the Line Length

In practice, we often encounter differences between the modeled line lengths and the actual line lengths. Normally, the modeled line lengths are based on the geographical routes, not on the actual lengths of the laid cable. The actual route has many small detours from the modeled route and those detours can add up to a significant percentage of the total cable length. This problem is slightly less severe, but still present, in overhead lines. Here, the sags between the poles are not accounted for.

The line between buses the 6320 and 671 is 1500 ft long. A $\pm 20\%$ interval accounts for ± 300 ft. Since the line has a three-phase configuration, the line impedance matrix has three self-impedances and three mutual impedances. The following subsections show the results for different measurement configurations.

4.2.6.1   Measurement Configuration PQ

Figures 4.6 and 4.7 show the interval of the magnitude and phase parts of the state vector, respectively, in terms of the $z_{6320-671}$, which represents an uncertain line length, for the SHGM and WLS estimators. It is clear that the local sensitivities of the two estimators are almost identical.

Figures 4.8 and 4.9 show the local sensitivity for the LAV estimator in comparison to the WLS. It is clear that the influence of the uncertain line length is much smaller when the LAV estimator is used.

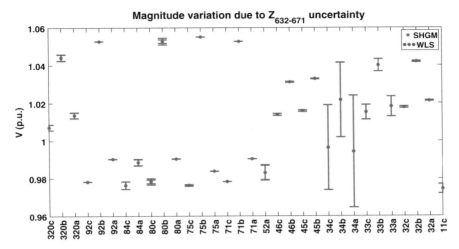

Fig. 4.6 Interval of the magnitude part of the state vector due to the uncertain $Z_{6320-671}$

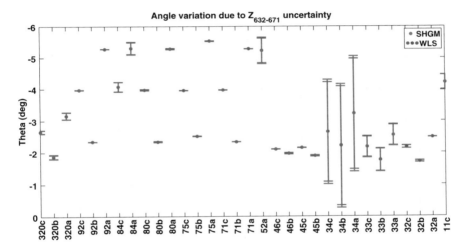

Fig. 4.7 Interval of the angle part of the state vector due to the uncertain $Z_{6320-671}$

Furthermore, the uncertainty interval is the largest on buses 633 and 634. The probable reason for this is that the series impedance of the transformer on branch $633 - 634$ is two orders of magnitude larger than the series impedance of any other branch. The uncertainty spreads across all the buses due to the measurement configuration and the nature of the power-injection measurement functions (Eqs. 2.32 and 2.33).

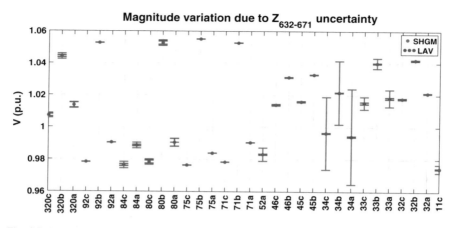

Fig. 4.8 Interval of the magnitude part of the state vector due to the uncertain $Z_{6320-671}$

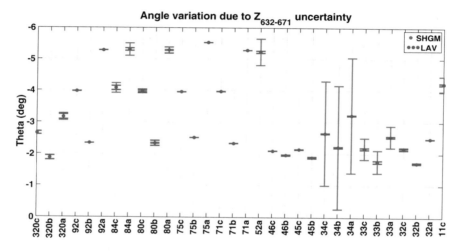

Fig. 4.9 Interval of the angle part of the state vector due to the uncertain $Z_{6320-671}$

4.2.6.2 Measurement Configuration PQ-PMU

In this measurement configuration, the effect of the uncertain line length on the estimated state variables is much smaller for the WLS and SHGM estimators and remains negligible for the LAV estimator as Figs. 4.10 and 4.11 show.

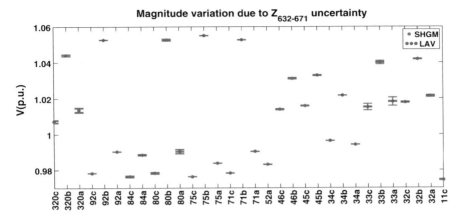

Fig. 4.10 Interval of the magnitude part of the state vector due to the uncertain $Z_{6320-671}$

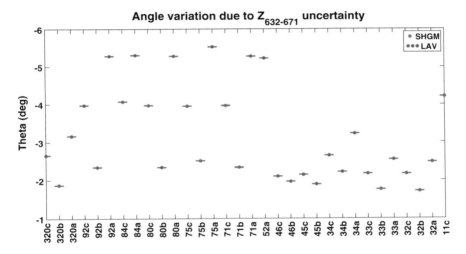

Fig. 4.11 Interval of the angle part of the state vector due to the uncertain $Z_{6320-671}$

4.2.6.3 Measurement Configuration PMU-PQ

The effect of the uncertain line length on the state vector in this configuration is even smaller than that in the previous two configurations, for all four estimators. Figures 4.12 and 4.13 depict the results for the WLS and LAV estimators. The two are virtually indistinguishable. We omitted the results for the SHGM, but the results are no different for this estimator.

The reason for such a small influence can be found in the measurement configuration in which the PMUs represent the majority of devices. The phase-angle and magnitude measurements alone form a linear state estimation problem, where the

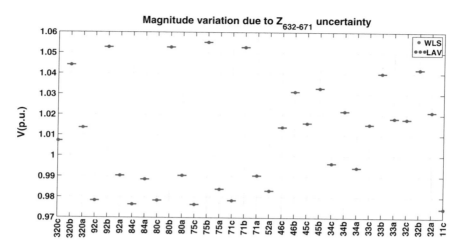

Fig. 4.12 Interval of the magnitude part of the state vector due to the uncertain $\mathbf{Z}_{6320-671}$

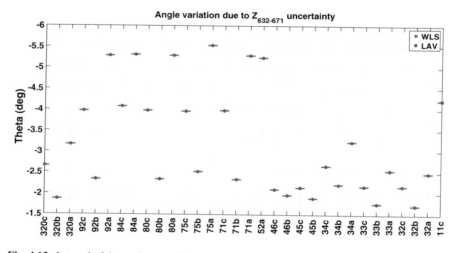

Fig. 4.13 Interval of the angle part of the state vector due to the uncertain $\mathbf{Z}_{6320-671}$

topology has no effect on the state variables. With the introduction of additional power-injection measurements, the topology has an effect on certain state variables, but since the measurements are only placed on a few buses and the influence of the topology uncertainty spreads in accordance with Eqs. 2.82 and 2.83, the majority of the buses remain isolated from the uncertainty influence.

4.2.7 Quality of the Measurements

Measurement devices are susceptible to several types of errors. The first type of measurement errors are small drifts of the measured value compared to the true value. For example, the time drifts of certain analog elements can cause errors in the measurement. The second type are gross measurement errors, e.g., a current transducer might start malfunctioning and the readings would be several orders of magnitude off the true value. In this study, we are interested in the first type of measurement errors, as the second type requires a different type of analysis.

For each of the measurement configurations and different estimators, the effect of uncertain measurements was computed. Different measurement weights determine the magnitude of the influence of a particular measurement in comparison to others, but to simplify the problem, the same weights were assigned to all the measurements.

4.2.7.1 Measurement Configuration PQ

The variation of the state vector for a ±20% uncertainty interval on power-injection measurements on bus 671 is shown in Figs. 4.14 and 4.15 for the WLS and SHGM estimators. It is clear that the estimators have a similar local sensitivity. Surprisingly, the LAV estimator has a much greater local sensitivity, as shown in Figs. 4.16 and 4.17.

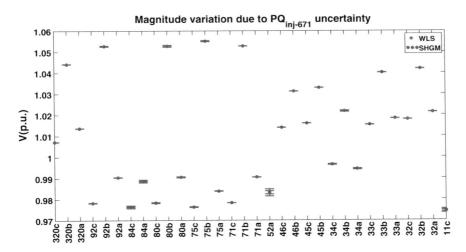

Fig. 4.14 Interval of the magnitude part of the state vector due to the uncertain $\mathbf{PQ}_{inj-671}$

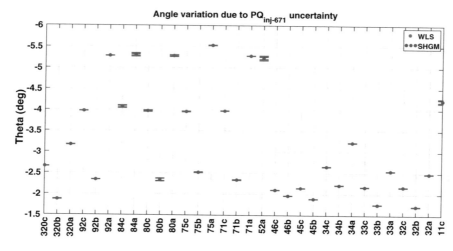

Fig. 4.15 Interval of the angle part of the state vector due to the uncertain $\mathbf{PQ}_{inj-671}$

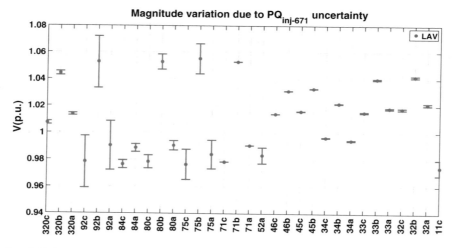

Fig. 4.16 Interval of the magnitude part of the state vector due to the uncertain $\mathbf{PQ}_{inj-671}$

4.2.7.2 Measurement Configuration PQ-PMU

The variation of the state vector for a ±20% uncertainty interval of the power-injection measurement on bus 671 is shown in Figs. 4.18 and 4.19. It is clear that the angle-measurement variation is barely observable in both the LAV and SHGM estimators. Interestingly, it can be observed that, similar to the PQ configuration, the LAV estimator has a greater local sensitivity with respect to the magnitude part than the SHGM. The WLS estimator behaves much like the SHGM estimator.

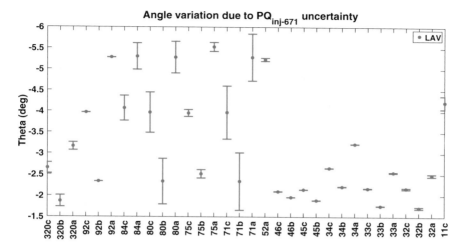

Fig. 4.17 Interval of the angle part of the state vector due to the uncertain $\mathbf{PQ}_{inj-671}$

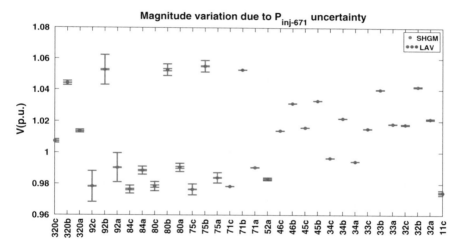

Fig. 4.18 Interval of the magnitude part of the state vector due to the uncertain $\mathbf{PQ}_{inj-671}$

4.2.7.3 Measurement Configuration PMU-PQ

Figures 4.20 and 4.21 show the state vector uncertainty interval for $\pm 0.01°$ uncertain phase measurements on bus 634. It is clear that the angle-measurement variation is almost fully translated into the state vector on buses 634 and 633, but remains isolated on the phase part of these two buses. Again, the reasons for this are to be found in the measurement configuration. The power-injection measurement on bus 634 causes a spread of the uncertainty onto the adjacent bus.

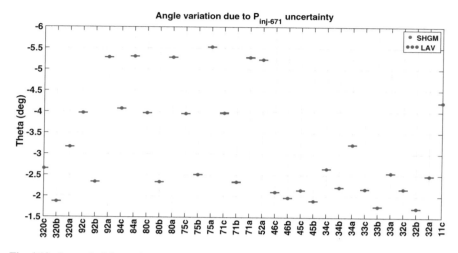

Fig. 4.19 Interval of the angle part of the state vector due to the uncertain $\mathbf{PQ}_{inj-671}$

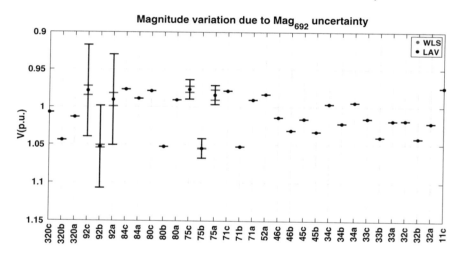

Fig. 4.20 Interval of the magnitude part of the state vector due to the uncertain $|\mathbf{V}|_{692}$

4.2.8 Discussion

In this chapter, we investigated the local sensitivities of estimated state variables with respect to the uncertain line lengths and inaccurate (pseudo-) measurements in a three-phase distribution network for different measurement configurations and different estimators. We selected and implemented the approach with a perturbation of the KKT conditions, which is agnostic to the choice of the estimator. The analysis was

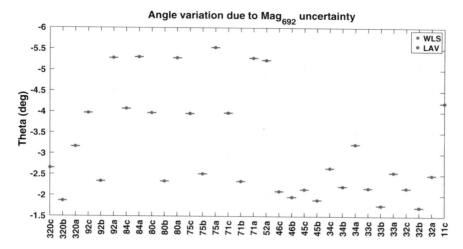

Fig. 4.21 Interval of the angle part of the state vector due to the uncertain $|\mathbf{V}|_{692}$

performed for a full three-phase branch-network model. The implemented estimators were LAV, WLS, and SHGM.

The results show that both the measurement configuration and the estimator choice have a large influence on the spread of the line impedance and measurement uncertainties across the network buses in the estimated state vector. The LAV estimator has a small local sensitivity with respect to the line impedances in comparison to the SHGM and WLS estimators. The opposite is the case for the measurement sensitivities, where the results show that the LAV estimator has the worst performance among the implemented examples. It can be concluded that the LAV estimator is suitable for networks where the line lengths are uncertain, but we have good information about measurement noise characteristics.

In terms of local sensitivities, the results show a very similar behavior for the SHGM and WLS estimators. However, among the two, only the SHGM estimator is considered robust in terms of gross measurement errors. Thus, the SHGM estimator based on projection statistics is suggested in cases where the line impedances are known with accuracy, but we are unsure about measurement noise characteristics.

Even though the influence of the line-impedance's uncertainty on the state vector estimation is smaller for the LAV estimator and for measurement configurations with a prevailing number of PMU measurements (in comparison to those with the prevailing number of power-injection measurements), the line impedances's uncertainty still influences the calculation of the derived variables (e.g., power flows). Thus, accurate knowledge of the line impedances is crucial for an accurate estimation of the state vector and its derived variables.

References

1. U. Kuhar, Three-Phase State Estimation in Power Distribution Systems. PhD thesis. Ljubljana, Slovenia: Jožef Stefan International Postgraduate School, 2018
2. U. Kuhar et al., The impact of model and measurement uncertainties on a state estimation in three-phase distribution networks. IEEE Trans. Smart Grid **10**(3), pp. 3301–3310 (2019) ISSN: 1949-3053, 1949-3061. https://doi.org/10.1109/TSG.2018.2823398
3. T.A. Stuart, C.J. Herczet, A sensitivity analysis of weighted least squares state estimation for power systems. IEEE Trans. Power Appar. Syst. **5**, 1696–1701 (1973)
4. D. Macii, G. Barchi, D. Petri, Uncertainty sensitivity analysis of WLS-based grid state estimators, in: *IEEE International Workshop on Applied Measurements for Power Systems (AMPS) Proceedings, 2014* (IEEE, 2014), pp. 1–6
5. A.K. Al-Othman, M.R. Irving, Uncertainty modelling in power system state estimation. IEE Proc.-Gener., Transm. Distrib. **152**(2), 233 (2005)
6. A.K. Al-Othman, M.R. Irving, Analysis of confidence bounds in power system state estimation with uncertainty in both measurements and parameters. Electr. Power Syst. Res. **76**(12), 1011–1018 (2006)
7. C. Rakpenthai, S. Uatrongjit, S. Premrudeepreechacharn, State estimation of power system considering network parameter uncertainty based on parametric interval linear systems. IEEE Trans. Power Syst. **27**(1), 305–313 (2012)
8. S. Chakrabarti, E. Kyriakides, PMU measurement uncertainty considerations in WLS state estimation. IEEE Trans. Power Syst. **24**(2), 1062–1071 (2009)
9. S. Chakrabarti, E. Kyriakides, M. Albu, Uncertainty in power system state variables obtained through synchronized measurements. IEEE Trans. Instrum. Meas. **58**(8), 2452–2458 (2009)
10. C. Muscas et al., Impact of different uncertainty sources on a three- phase state estimator for distribution networks. IEEE Trans. Instrum. Meas. **63**(9), 2200–2209 (2014)
11. R. Minguez, A.J. Conejo, State estimation sensitivity analysis. IEEE Trans. Power Syst. **22**(3), 1080–1091 (2007)
12. E. Castillo et al., Perturbation approach to sensitivity analysis in mathematical programming. J. Optim. Theory Appl. **128**(1), 49–74 (2006)
13. E. Castillo, *Building and Solving Mathematical Programming Models in Engineering and Science* (Wiley, 2002) ISBN: 978-0-471-15043-5
14. W.H. Kersting, *Distribution System Modeling and Analysis. The Electric Power Engineering Series*. (CRC Press, Boca Raton, 2002) ISBN: 0-8493-0812- 7
15. E. Caro, A.J. Conejo, R. Minguez, A sensitivity analysis method to compute the residual covariance matrix. Electr. Power Syst. Res. **81**(5), 1071–1078 (2011) ISSN: 03787796. https://doi.org/10.1016/j.epsr.2010.12.007
16. W.H. Kersting, Radial distribution test feeders, in *Power Engineering Society Winter Meeting, 2001. IEEE*, vol. 2 (IEEE, 2001), pp. 908–912. Accessed 16 Oct 2015
17. U. Kuhar, G. Kosec, A. Svigelj, *Measurement Noise Propagation in Distribution-system State Estimation*. (IEEE, 2017), pp. 1049–1054. ISBN: 978-953-233-090-8. https://doi.org/10.23919/MIPRO.20177973579. url: http://ieeexplore.ieee.org/document/7973579/ Accessed 23 Jul 2018

Chapter 5
Final Remarks and Conclusions

5.1 Book Recapitulation

Recent changes in power-distribution systems will require close to real-time observation and control of the system to ensure a safe and reliable operation in the future. DSSE presents a crucial part in ensuring the observability. It acts like a multi-variate filter of the measurements deployed in the field, capable of outputting the most likely system state. This, in turn, is the information that present and future control functions leverage to ensure a safe and reliable operation of the system [1, 2].

In this book, the main goal was to develop, implement, and thoroughly evaluate a three-phase DSSE. A unified three-phase branch model that allows the unified modeling of conductors, transformers, tap changers, and voltage regulators was presented [3]. The development of model parameters for all the components was reviewed. The problem of a statistical state estimation, and how it applies to the estimation of the power-distribution's system state was also reviewed. A three-phase DSSE that can incorporate heterogeneous measurement types was developed. Several different state estimation algorithms, robust and non-robust, were reviewed and compared based on their statistical efficiency and computational speed. The developed DSSE with all the implemented estimation norms (WLS, SHGM, and LAV) was evaluated for the sensitivity of the state variables to model and measurement uncertainties. The analysis included different measurement configurations.

The results presented in this book are expected to contribute to further research and the development of methods, models, and procedures. The developed unified three-phase branch model eases the modeling of the distribution power network. This book includes examples of model applications to conductor sections, transformers, tap changers, and voltage regulators.

The presented sensitivity analysis enables a sensitivity evaluation of the chosen state estimation algorithms and the measurement configuration. It was shown that this is highly desirable, as it enables an informed choice about the most suitable state estimation algorithm based on the configuration of measurements and the accuracy

U. Kuhar et al., *Observability of Power-Distribution Systems*,
SpringerBriefs in Applied Sciences and Technology,
https://doi.org/10.1007/978-3-030-39476-9_5

of the system model [4]. The presented lower and upper interval bounds give a clear indication of the performance of the state estimation system that can be expected with the given accuracy of the model parameters and measurements.

5.2 Future Work

The work presented in this book could have several possible research and development directions:

- A large number of buses and, as a consequence, measurement devices will create enormous quantities of data that will need to be stored for future manipulation. Technologies that would enable the efficient storage and fast recall of measurements and state estimation results will need to be developed. Some research is already underway with the Berkeley Tree Database project [5].
- Further research in current and new applications that leverage state-estimation results will be necessary. Applications such as

  - estimation of impedances and model validation,
  - detection of oscillations,
  - islanding detection and resynchronization,
  - topology detection and the real-time estimation of loads,
  - fault-detection localization and classification,

  and many others, create true customer value in terms of better utilization of the available resources, delayed investments into infrastructure, and greater system reliability.

References

1. U. Kuhar, Three-phase state estimation in power distribution systems. PhD thesis, Jožef Stefan International Postgraduate School, Ljubljana, Slovenia, 2018
2. J.J. Nielsen et al., Secure real-time monitoring and management of smart distribution grid using shared cellular networks. IEEE Wirel. Commun. **24**(2), 10–17 (2017). ISSN: 1536-1284. https://doi.org/10.1109/MWC.2017.1600252
3. U. Kuhar et al., A unified three-phase branch model for a distribution-system state estimation, in *PES Innovative Smart Grid Technologies Conference Europe (ISGT-Europe), 2016, IEEE* (IEEE, 2016), pp. 1–6
4. U. Kuhar et al., The impact of model and measurement uncertainties on a state estimation in three-phase distribution networks. IEEE Trans. Smart Grid **10**(3), 3301–3310 (2019). ISSN: 1949-3053, 1949-3061. https://doi.org/10.1109/TSG.2018.2823398.
5. Website. Berkeley tree database (2018), http://btrdb.io/. Accessed 25 Nov 2019

Appendix A
IEEE Distribution Test Feeders

A.1 The 13-Bus Feeder

The topology of the IEEE 13-bus feeder is included here. Figure A.1 depicts the one-line diagram of the feeder. Labels on branches represent phasing sequences, and labels near spot loads represent load type and its connection (D—delta connection, Y—wye connection, PQ—constant power injection load, Z—constant impedance load, I—constant current load).

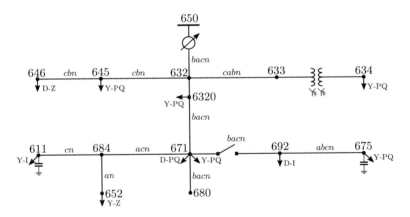

Fig. A.1 IEEE 13-bus feeder

U. Kuhar et al., *Observability of Power-Distribution Systems*,
SpringerBriefs in Applied Sciences and Technology,
https://doi.org/10.1007/978-3-030-39476-9

Table A.1 lists the topology data.

Table A.1 IEEE 13—bus topology data

Node A	Node B	Length (ft.)	Type	Phasing	Phase No.	Neutral No.	Spacing ID
632	645	500	OHL	CBN	8	8	505
632	633	500	OHL	CABN	6	6	500
633	634	0	Transf.				
645	646	300	OHL	CBN	8	8	505
650	632	2000	OHL	BACN	2	6	500
684	652	800	UGC				
632	671	2000	OHL	BACN	2	6	500
671	684	300	OHL	ACN	8	8	505
671	680	1000	OHL	BACN	2	6	500
671	692	0	Switch				
684	611	300	OHL	CN	8	8	510
692	675	500	UGC				

Table A.2 lists the spot loads on the feeder.

Table A.2 IEEE 13—spot loads

Node	Load type	Ph1 (kW)	Ph1 (kVAr)	Ph2 (kW)	Ph2 (kVAr)	Ph3 (kW)	Ph3 (kVAr)
634	Y-PQ	160	110	120	90	120	90
645	Y-PQ	0	0	170	125	0	0
646	D-Z	0	0	230	132	0	0
652	Y-Z	128	86	0	0	0	0
671	D-PQ	385	220	385	220	385	220
675	Y-PQ	485	190	68	60	290	212
692	D-I	0	0	0	0	170	151
611	Y-I	0	0	0	0	170	80

Table A.3 lists the distributed loads on the feeder.

Table A.3 IEEE 13—distributed loads

Node A	Node B	Type	Ph1	Ph1	Ph2	Ph2	Ph3	Ph3
632	671	Y-PQ	17 kW	10 kVAr	66 kW	38 kVAr	117 kW	68 kVAr

Table A.4 lists the shunt capacitors on the feeder.

Table A.4 IEEE 13—shunt capacitors

Node	Connection	Ph1 (kVAr)	Ph2 (kVAr)	Ph3 (kVAr)
675	Y	200	200	200
611	Y	0	0	100

Table A.5 lists the transformers on the feeder.

Table A.5 IEEE 13—transformer

No.	Node A	Node B	Power	P. vol.	P. conn.	S. vol.	S. conn.	R	X
1	633	634	500 kVA	4.16 kV	Yg	0.48 kV	Yg	1.1%	2%

Table A.6 lists the voltage regulators on the feeder.

Table A.6 IEEE 13—voltage regulators

Property	Value 1	Value 2	Value 3
Regulator ID	1		
Line segment	650	632	
Location	650		
Phases	A-B-C		
Connection	3-Ph, L-G		
Monitoring phase	A-B-C		
Bandwidth	2	Volts	
Potential transf. ratio	20		
Primary current transf. ratio	700		
Compensator settings	Ph-A	Ph-B	Ph-C
R setting	3	3	9
X setting	9	9	9
Voltage level	122	122	122

A.2   The 123-Bus Feeder

The topology of the IEEE 123-bus feeder is included here. Figure A.2 depicts the one-line diagram of the feeder (Table A.7).

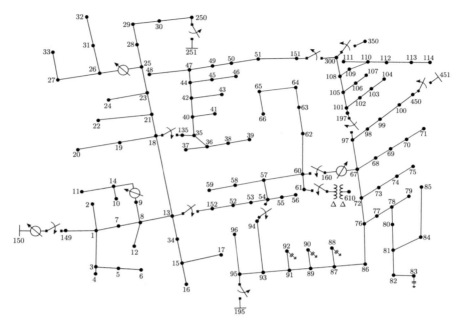

Fig. A.2 IEEE 123-bus feeder

Table A.7 IEEE 123—bus topology data

Node A	Node B	Length (ft.)	Type	Phasing	Phase No.	Neutral No.	Spacing ID
1	2	175	OHL	BN	8	8	510
1	3	250	OHL	CN	8	8	510
1	7	300	OHL	ABCN	5	6	500
3	4	200	OHL	CN	8	8	510
3	5	325	OHL	CN	8	8	510
5	6	250	OHL	CN	8	8	510
7	8	200	OHL	ABCN	5	6	500
8	12	225	OHL	BN	8	8	510
8	9	225	OHL	AN	8	8	510
8	13	300	OHL	ABCN	5	6	500
9	14	425	OHL	AN	8	8	510
13	34	150	OHL	CN	8	8	510
13	18	825	OHL	CABN	5	6	500
13	152	0	Switch				
18	135	0	Switch				

(continued)

Table A.7 (continued)

Node A	Node B	Length (ft.)	Type	Phasing	Phase No.	Neutral No.	Spacing ID
14	11	250	OHL	AN	8	8	510
14	10	250	OHL	AN	8	8	510
15	16	375	OHL	CN	8	8	510
15	17	350	OHL	CN	8	8	510
18	21	300	OHL	CABN	5	6	500
19	20	325	OHL	AN	8	8	510
21	22	525	OHL	BN	8	8	510
21	23	250	OHL	CABN	5	6	500
23	24	550	OHL	CN	8	8	510
23	25	275	OHL	CABN	5	6	500
25	26	350	OHL	ACN	5	6	505
26	31	225	OHL	CN	8	8	510
27	33	500	OHL	AN	8	8	510
28	29	300	OHL	CABN	5	6	500
29	30	350	OHL	CABN	5	6	500
30	250	200	OHL	CABN	5	6	500
31	32	300	OHL	CN	8	8	510
34	15	100	OHL	CN	8	8	510
35	36	650	OHL	ABN	5	6	505
35	40	250	OHL	ABCN	5	6	500
36	37	300	OHL	AN	8	8	510
36	38	250	OHL	BN	8	8	510
38	39	325	OHL	BN	8	8	510
40	41	325	OHL	CN	8	8	510
40	42	250	OHL	ABCN	5	6	500
42	43	500	OHL	BN	8	8	510
42	44	200	OHL	ABCN	5	6	500
36	37	300	OHL	AN	8	8	510
44	45	300	OHL	AN	8	8	510
44	47	250	OHL	ABCN	5	6	500
45	46	300	OHL	AN	8	8	510
47	48	150	OHL	CBAN	5	6	500
47	49	250	OHL	CBAN	5	6	500
49	50	250	OHL	CBAN	5	6	500
50	51	250	OHL	CBAN	5	6	500
51	151	500	OHL	CBAN	5	6	500

(continued)

Table A.7 (continued)

Node A	Node B	Length (ft.)	Type	Phasing	Phase No.	Neutral No.	Spacing ID
52	53	200	OHL	ABCN	5	6	500
53	54	125	OHL	ABCN	5	6	500
54	55	275	OHL	ABCN	5	6	500
54	57	350	OHL	BCAN	5	6	500
55	56	275	OHL	ABCN	5	6	500
57	58	250	OHL	BN	8	8	510
57	60	750	OHL	BCAN	5	6	500
58	59	250	OHL	BN	8	8	510
60	61	550	OHL	BACN	5	6	500
60	62	250	UGC	ABC	2		515
60	160	0	Switch				
61	610	0	Switch				
61	610	0	XFM-1				
62	63	175	UGC	ABC	2		515
63	64	350	UGC	ABC	2		515
64	65	425	UGC	ABC	2		515
65	66	325	UGC	ABC	2		515
67	68	200	OHL	AN	8	8	510
67	72	275	OHL	BCAN	5	6	500
67	97	250	OHL	BCAN	5	6	500
68	69	275	OHL	AN	8	8	510
69	70	325	OHL	AN	8	8	510
70	71	275	OHL	AN	8	8	510
72	73	275	OHL	CN	8	8	510
72	76	200	OHL	BCAN	5	6	500
73	74	350	OHL	AN	8	8	510
74	75	400	OHL	AN	8	8	510
76	77	200	OHL	ACBN	5	6	500
76	86	700	OHL	BCAN	5	6	500
77	78	100	OHL	ACBN	5	6	500
78	79	225	OHL	ACBN	5	6	500
78	80	475	OHL	ACBN	5	6	500
80	81	475	OHL	ACBN	5	6	500
81	82	250	OHL	ACBN	5	6	500
81	84	675	OHL	CN	8	8	510
82	83	250	OHL	ACBN	5	6	500
84	85	475	OHL	CN	8	8	510
86	87	450	OHL	ACBN	5	6	500
87	88	175	OHL	AN	8	8	510

(continued)

Table A.7 (continued)

Node A	Node B	Length (ft.)	Type	Phasing	Phase No.	Neutral No.	Spacing ID
87	89	275	OHL	ACBN	5	6	500
89	90	225	OHL	BN	8	8	510
89	91	225	OHL	ACBN	5	6	500
91	92	300	OHL	CN	8	8	510
91	93	225	OHL	ACBN	5	6	500
93	94	275	OHL	AN	8	8	510
93	95	300	OHL	ACBN	5	6	500
95	96	200	OHL	BN	8	8	510
97	98	275	OHL	BCAN	5	6	500
97	197	0	Switch				
98	99	550	OHL	BCAN	5	6	500
99	100	300	OHL	BCAN	5	6	500
100	450	800	OHL	BCAN	5	6	500
101	102	225	OHL	CN	8	8	510
101	105	275	OHL	BCAN	5	6	500
102	103	325	OHL	CN	8	8	510
103	104	700	OHL	CN	8	8	510
105	106	225	OHL	BN	8	8	510
105	108	325	OHL	BCAN	5	6	500
106	107	575	OHL	BN	8	8	510
108	109	450	OHL	AN	8	8	510
108	300	1000	OHL	BCAN	5	6	500
109	110	300	OHL	AN	8	8	510
110	111	575	OHL	AN	8	8	510
110	112	125	OHL	AN	8	8	510
112	113	525	OHL	AN	8	8	510
113	114	325	OHL	AN	8	8	510
135	35	375	OHL	CBAN	5	6	500
149	1	400	OHL	ABCN	5	6	500
150	149	0	Switch				
152	52	400	OHL	ABCN	5	6	500
160	67	350	OHL	ACBN	5	6	500
197	101	250	OHL	BCAN	5	6	500
250	251	0	Switch				
450	451	0	Switch				
54	94	0	Switch				
151	300	0	Switch				
300	350	0	Switch				

Table A.8 lists the spot loads on the feeder.

Table A.8 IEEE 123—spot loads

Node	Load type	Ph1 (kW)	Ph1 (kVAr)	Ph2 (kW)	Ph2 (kVAr)	Ph3 (kW)	Ph3 (kVAr)
1	Y-PQ	40	20	0	0	0	0
2	Y-PQ	0	0	20	10	0	0
4	Y-PQ	0	0	0	0	40	20
5	Y-I	0	0	0	0	20	10
6	Y-Z	0	0	0	0	40	20
7	Y-PQ	20	10	0	0	0	0
9	Y-PQ	40	20	0	0	0	0
10	Y-I	20	10	0	0	0	0
11	Y-Z	40	20	0	0	0	0
12	Y-PQ	0	0	20	10	0	0
16	Y-PQ	0	0	0	0	40	20
17	Y-PQ	0	0	0	0	20	10
19	Y-PQ	40	20	0	0	0	0
20	Y-I	40	20	0	0	0	0
22	Y-Z	0	0	40	20	0	0
24	Y-PQ	0	0	0	0	40	20
28	Y-I	40	20	0	0	0	0
29	Y-Z	40	20	0	0	0	0
30	Y-PQ	0	0	0	0	40	20
31	Y-PQ	0	0	0	0	20	10
32	Y-PQ	0	0	0	0	20	10
33	Y-I	40	20	0	0	0	0
34	Y-Z	0	0	0	0	40	20
35	D-PQ	40	20	0	0	0	0
37	Y-Z	40	20	0	0	0	0
38	Y-I	0	0	20	10	0	0
39	Y-PQ	0	0	20	10	0	0
41	Y-PQ	0	0	0	0	20	10
42	Y-PQ	20	10	0	0	0	0
43	Y-Z	0	0	40	20	0	0
45	Y-I	20	10	0	0	0	0
46	Y-PQ	20	10	0	0	0	0

(continued)

Table A.8 (continued)

Node	Load type	Ph1 (kW)	Ph1 (kVAr)	Ph2 (kW)	Ph2 (kVAr)	Ph3 (kW)	Ph3 (kVAr)
47	Y-I	35	25	35	25	35	25
48	Y-Z	70	50	70	50	70	50
49	Y-PQ	35	25	70	50	35	20
50	Y-PQ	0	0	0	0	40	20
51	Y-PQ	20	10	0	0	0	0
52	Y-PQ	40	20	0	0	0	0
53	Y-PQ	40	20	0	0	0	0
55	Y-Z	20	10	0	0	0	0
56	Y-PQ	0	0	20	10	0	0
58	Y-I	0	0	20	10	0	0
59	Y-PQ	0	0	20	10	0	0
60	Y-PQ	20	10	0	0	0	0
62	Y-Z	0	0	0	0	40	20
63	Y-PQ	40	20	0	0	0	0
64	Y-I	0	0	75	35	0	0
65	D-Z	35	25	35	25	70	50
66	Y-PQ	0	0	0	0	75	35
68	Y-PQ	20	10	0	0	0	0
69	Y-PQ	40	20	0	0	0	0
70	Y-PQ	20	10	0	0	0	0
71	Y-PQ	40	20	0	0	0	0
73	Y-PQ	0	0	0	0	40	20
74	Y-Z	0	0	0	0	40	20
75	Y-PQ	0	0	0	0	40	20
76	D-I	105	80	70	50	70	50
77	Y-PQ	0	0	40	20	0	0
79	Y-Z	40	20	0	0	0	0
80	Y-PQ	0	0	40	20	0	0
82	Y-PQ	40	20	0	0	0	0
83	Y-PQ	0	0	0	0	20	10
84	Y-PQ	0	0	0	0	20	10
85	Y-PQ	0	0	0	0	40	20
86	Y-PQ	0	0	20	10	0	0
87	Y-PQ	0	0	40	20	0	0
88	Y-PQ	40	20	0	0	0	0
90	Y-I	0	0	40	20	0	0

(continued)

Table A.8 (continued)

Node	Load type	Ph1 (kW)	Ph1 (kVAr)	Ph2 (kW)	Ph2 (kVAr)	Ph3 (kW)	Ph3 (kVAr)
92	Y-PQ	0	0	0	0	40	20
94	Y-PQ	40	20	0	0	0	0
95	Y-PQ	0	0	20	10	0	0
96	Y-PQ	0	0	20	10	0	0
98	Y-PQ	40	20	0	0	0	0
99	Y-PQ	0	0	40	20	0	0
100	Y-Z	0	0	0	0	40	20
102	Y-PQ	0	0	0	0	20	10
103	Y-PQ	0	0	0	0	40	20
104	Y-PQ	0	0	0	0	40	20
106	Y-PQ	0	0	40	20	0	0
107	Y-PQ	0	0	40	20	0	0
109	Y-PQ	40	20	0	0	0	0
111	Y-PQ	20	10	0	0	0	0
112	Y-I	20	10	0	0	0	0
113	Y-Z	40	20	0	0	0	0
114	Y-PQ	20	10	0	0	0	0

Table A.9 lists the shunt capacitors on the feeder.

Table A.9 IEEE 123—shunt capacitors

Node	Connection	Ph1 (kVAr)	Ph2 (kVAr)	Ph3 (kVAr)
83	Y	200	200	200
88	Y	50	0	0
90	Y	0	50	0
92	Y	0	0	50

Table A.10 lists the transformers on the feeder.

Table A.10 IEEE 13—transformer

No.	Node A	Node B	Power (kVA)	P. vol. (kV)	P. conn.	S. vol. (kV)	S. conn.	R (%)	X (%)
1		150	5000	115	D	4.16	Yg	1	8
2	61	610	150	4.16	D	0.48	D	1.27	2.72

Table A.11 lists the voltage regulators on the feeder.

Table A.11 IEEE 123—voltage regulators

Property	Value 1	Value 2	Value 3
Regulator ID	1		
Line segment	150	149	
Location	150		
Phases	A-B-C		
Connection	3-Ph, L-G		
Monitoring phase	A		
Bandwidth	2	Volts	
Potential transf. ratio	20		
Primary current transf. ratio	700		
Compensator settings	Ph-A		
R setting	3		
X setting	7, 5		
Voltage level	120		
Regulator ID	2		
Line segment	9	14	
Location	9		
Phases	A		
Connection	1-Ph, L-G		
Monitoring phase	A		
Bandwidth	2	Volts	
Potential transf. ratio	20		
Primary current transf. ratio	50		
Compensator settings	Ph-A		
R setting	0, 4		
X setting	0, 4		
Voltage level	120		
Regulator ID	3		
Line segment	25	26	
Location	25		
Phases	A-C		
Connection	2-Ph, L-G		
Monitoring phase	A-C		
Bandwidth	1	Volts	
Potential transf. ratio	20		
Primary current transf. ratio	50		

(continued)

Table A.11 (continued)

Property	Value 1	Value 2	Value 3
Compensator settings	Ph-A	Ph-C	
R setting	0,4	0,4	
X setting	0,4	0,4	
Voltage level	120	120	
Regulator ID	4		
Line segment	160	67	
Location	160		
Phases	A-B-C		
Connection	3-Ph, L-G		
Monitoring phase	A-B-C		
Bandwidth	2	Volts	
Potential transf. ratio	20		
Primary current transf. ratio	300		
Compensator settings	Ph-A	Ph-B	Ph-C
R setting	0,6	1,4	0,2
X setting	1,3	2,6	1,4
Voltage level	124	124	124

Table A.12 lists the mixed measurement configuration for validation.

Table A.12 Mixed measurement configuration for validation

Meas. confg.	1		2	3	4	5	6	7
Bus/Branch	PMU	PQ	PQ	PQ	PQ	PQ	PQ	PQ
1−149	•							•
2−1	•							•
3−1	•							•
4−3	•							•
5−3	•							•
6−5	•							•
7−1	•							•
8−7	•							•
9−8	•							•
10−14	•							•
11−14	•							•
12−8	•						•	

(continued)

Table A.12 (continued)

Meas. confg.	1		2	3	4	5	6	7
Bus/Branch	PMU	PQ	PQ	PQ	PQ	PQ	PQ	PQ
13–8	●						●	
14–9	●						●	
15–34	●						●	
16–15	●						●	
17–15	●						●	
18–13	●						●	
19–18	●						●	
20–19	●						●	
21–18	●						●	
22–21	●						●	
23–21	●					●		
24–23	●					●		
25–23	●					●		
26–25	●					●		
27–26	●					●		
28–25	●					●		
29–28	●					●		
30–29	●					●		
31–26	●					●		
32–31	●					●		
33–27	●					●		
34–13	●				●			
35–135	●				●			
36–35	●				●			
37–36	●				●			
38–36	●				●			
39–38	●				●			
40–35	●				●			
41–40	●				●			
42–40	●				●			
43–42	●				●			
44–42	●				●			
45–44	●			●				
46–45	●			●				
47–44	●			●				
48–47	●			●				
49–47	●			●				

(continued)

Table A.12 (continued)

Meas. confg.	1		2	3	4	5	6	7
Bus/Branch	PMU	PQ	PQ	PQ	PQ	PQ	PQ	PQ
50−49	•			•				
51−50	•			•				
52−152	•			•				
53−52	•			•				
54−53	•			•				
55−54	•			•				
56−55	•		•					
57−54	•		•					
58−57	•		•					
59−58	•		•					
60−57	•		•					
61−60	•		•					
62−60	•		•					
63−62	•		•					
64−63	•		•					
65−64	•		•					
66−65	•		•					
67−160	•		•					
68−67	•	•						
69−68	•	•						
70−69	•	•						
71−70	•	•						
72−67	•	•						
73−72	•	•						
74−73	•	•						
75−74	•	•						
76−72	•	•						
77−76	•	•						
78−77	•	•						
79−78	•	•						
80−78	•	•						
81−80	•	•						
82−81	•	•						
83−82	•	•						
84−81	•	•						
85−84	•	•						
86−76	•	•						
87−89	•	•						

(continued)

Table A.12 (continued)

Meas. confg.	1		2	3	4	5	6	7
Bus/Branch	PMU	PQ	PQ	PQ	PQ	PQ	PQ	PQ
88−87	●	●						
89−87	●	●						
90−89	●	●						
91−89	●	●						
92−91	●	●						
93−91	●	●						
94−93	●	●						
95−93	●	●						
96−95	●	●						
97−67	●	●						
98−97	●	●						
99−98	●	●						
100−99	●	●						
101−197	●	●						
102−101	●	●						
103−102	●	●						
104−103	●	●						
105−101	●	●						
106−105	●	●						
107−106	●	●						
108−105	●	●						
109−108	●	●						
110−109	●	●						
111−110	●	●						
112−110	●	●						
113−112	●	●						
114−113	●	●						
135−18	●	●						
149−150	●	●						
151−51	●	●						
152−13	●	●						
160−60	●	●						
197−197	●	●						
250−30	●	●						
300−108	●	●						
450−100	●	●						
610−61	●	●						

Appendix B
Conductor Data

B.1 Conductor Data

Where conductor types are AA—All Aluminum, ACSR—Aluminum Conductor Steel Reinforced, CU—Copper, and resistance is specified for 60 Hz and 50 °C (Table B.1).

Table B.1 Conductor Data

No.	Conductor size		Stranding	Type	Resistivity	Diameter	GMR	Rating
	Circular miles	AWG			Ω/mile	Inch	Feet	Amperes
1	1000000	10.7/0		AA	0.105	1.15	0.0368	698
2	556500	8.17/0	26/7	ACSR	0.186	0.927	0.0311	730
3	500000	7.7/0		AA	0.206	0.813	0.026	483
4	250000	4.7/0		AA	0.41	0.567	0.0171	329
5	336400	6/0	26/7	ACSR	0.306	0.721	0.0244	530
6	211566	4/0	6/1	ACSR	0.592	0.563	0.00814	340
7	133056	2/0		AA	0.769	0.414	0.0125	230
8	105518	1/0		ACSR	1.12	0.398	0.00446	230
9	105518	1/0		AA	0.97	0.368	0.0111	202
10	105518	1/0		CU	0.607	0.368	0.01113	310
11	66361	2		AA	1.54	0.292	0.00883	156
12	66361	2	6/1	ACSR	1.69	0.316	0.00418	180
13	41735	4	6/1	ACSR	2.55	0.257	0.00452	140
14	10382	10		CU	5.903	0.102	0.00331	80
15	6529	12		CU	9.375	0.081	0.00262	75
16	208.080	14		CU	14.872	0.064	0.00208	20

© The Author(s), under exclusive license to Springer Nature Switzerland AG 2020
U. Kuhar et al., *Observability of Power-Distribution Systems*,
SpringerBriefs in Applied Sciences and Technology,
https://doi.org/10.1007/978-3-030-39476-9

B.2　Underground Cables

See the Table B.2

Table B.2　Underground concentric neutral all aluminum cable

	Concentric neutral 15kV All aluminum cable—1/3 neutral										
No.	Phase conductor size		Number of strands	Diameter over insulation	Diameter over screen	Outside diameter	Neutral conductor size		Number of strands	Neutral material	Ampacity
	Circular miles	AWG		Inch	Inch	Inch	Circular miles	AWG			
1	66361	2	7	0.78	0.85	0.98	208.080	14	6	CU	135
2	105518	1/0	19	0.85	0.93	1.06	208.080	14	6	CU	175
3	133056	2/0	19	0.9	0.97	1.1	208.080	14	7	CU	200
4	250000	4.7/0	37	1.06	1.16	1.29	208.080	14	13	CU	260
5	500000	7.7/0	37	1.29	1.39	1.59	6529	12	16	CU	385
6	1000000	10.7/0	61	1.64	1.77	1.98	10382	10	20	CU	550

Printed in the United States
By Bookmasters